AF588249

Neuroanatomy and Neurophysiology of the Larynx

Yasuo Hisa
Editor

Neuroanatomy and Neurophysiology of the Larynx

Springer

Editor
Yasuo Hisa
Department of Otolaryngology-Head and Neck Surgery
Kyoto Prefectural University of Medicine
Kyoto
Japan

ISBN 978-4-431-55749-4 ISBN 978-4-431-55750-0 (eBook)
DOI 10.1007/978-4-431-55750-0

Library of Congress Control Number: 2016956573

Printed on acid-free paper

This Springer imprint is published by Springer Nature
The registered company is Springer Japan KK

Foreword

I am pleased that Dr. Hisa has privileged me with being able to write the foreword for his textbook. Since much of the work in devising new therapies for patients with difficult laryngeal problems has involved the translational application of the microneuroanatomy of the larynx, this text should serve as a primary reference guide for all laryngologists.

Moreover, Dr. Hisa has been able to follow the neural organization of motor, sensory, and autonomic connections back to the brainstem, permitting even deeper insight into the organizational structure and control of the larynx. Recent ongoing studies have suggested that a number of the laryngeal disorders affecting patients may be the result of abnormal changes in the sensory system. For example, spasmodic dysphonia and paradoxical vocal cord motion demonstrate abnormal sensory feedback control of the larynx. This basic work provides the foundation upon which hopefully to understand and manage these types of conditions more effectively. I most heartily recommend this text to all clinician scientists interested in central control of the larynx in speech, respiration, and deglutition.

Gerald S. Berke, M.D.
Department of Head and Neck Surgery
UCLA,
Los Angeles, CA, USA

Preface

From the days of Galen in the second century A.D., elucidation of the neural control system of the larynx has been the target of many researchers, and many different methods have been employed. Advances in electrophysiological methods such as electromyography have made revolutionary findings possible in basic research of the larynx. Many mysteries have also been cleared up through painstaking accumulation of fine morphological data. However, interest in basic laryngology seems to have waned to some extent in recent days, and I am sorry to see that the precious work of our predecessors is on the verge of being forgotten.

My interest was first drawn to neuroscience of the larynx by a textbook description of the autonomic nerve fibers to the larynx reaching the larynx along the laryngeal arteries. In further reading, I found that there was no study that provided evidence for this. I did my own research on this topic and found that the autonomic nerve fibers do not follow the arteries, but reach the larynx through the superior and inferior laryngeal nerves. After this episode, I continued research in neuroanatomy of the larynx, and in time many colleagues of the Department of Otolaryngology–Head and Neck Surgery of the Kyoto Prefectural University of Medicine joined me, and together we have been able to broaden our understanding of the peripheral nervous system of the larynx and even to show how biological clock genes participate in laryngeal functional control. Electrophysiological studies provide not only simple electromyographic information on laryngeal muscles, but now have begun to produce accurate single-cell information on neurons belonging to the central complexes controlling respiration and deglutition. This book summarizes these developments in research on the larynx. I would like to express my gratitude to my fellow researchers who joined me in my researches and who also devoted extensive efforts to the creation of this book.

It is rare for a dedicated researcher to work exclusively on a specific field in otorhinolaryngology, much less laryngology. We have continued our basic research on the laryngeal innervation system despite the limitations in time imposed by clinical duties. It is my sincere wish that young researchers will follow in our steps and further develop the heritage of basic research in laryngology.

Finally, I would like to express my profound appreciation and gratitude to the late Professor Osamu Mizukoshi and to Professor Yasuhiko Ibata, the two people who opened the path to basic research in laryngology for me.

I dedicate this book to my wife, Yuko.

Yasuo Hisa, M.D., Ph.D.
Kyoto, Japan

Contents

Receptors and Nerve Endings

Sensory Receptors and Nerve Endings

Takeshi Nishio, Shinobu Koike, Hiroyuki Okano, and Yasuo Hisa

T. Nishio • S. Koike • H. Okano • Y. Hisa (✉)
Department of Otolaryngology-Head and Neck Surgery, Kyoto Prefectural University of Medicine, Kawaramachi-Hirokoji, Kamigyo-ku, Kyoto 602-8566, Japan
e-mail: nishio-t@koto.kpu-m.ac.jp; yhisa@koto.kpu-m.ac.jp

Y. Hisa (ed.), *Neuroanatomy and Neurophysiology of the Larynx*, DOI 10.1007/978-4-431-55750-0_1

1.1 Sensory Receptors

1.1.1 Introduction

The sensory senses include vision, hearing, olfaction, gustation, and general senses (touch, pressure, and proprioception). There are still many problems to be solved which neural system elicits the complex perception and mind with it. In order to understand the sensory neural systems, it is indispensable to elucidate the mechanism of neural receptors.

1.1.2 Classifications of Sensory Receptor

1.1.2.1 Classification by Construction

(a) Neuroepithelial receptors

There are neurons of which cell bodies exist in sensory epithelium on the surface of body, and these neurons directly convey information to the central nervous system. In the mammal, these neurons can be found only in the olfactory organs.

(b) Epithelial receptors

Some nonneural epithelial cells play a role of function as receptors. Taste buds or hair cells of inner ear correspond to them.

(c) Neuronal receptors

This type of reception cells is called the primary sensory neuron. All of receptors regarding superficial perception and proprioception belong in this group.

1.1.2.2 Classification by Stimulus Type Detected

(a) Mechanoreceptors

They generate nerve impulses when they are deformed by mechanical forces such as touch, pressure, vibration, stretch, and itch.

(b) Nociceptors

They respond to potentially damaging stimuli that result in pain.

(c) Chemoreceptors

They respond to chemicals in solution.

(d) Photoreceptors

Such as those of the retina of the eye, they respond to light energy.

(e) Thermoreceptors

They are sensitive to temperature changes.

1.1.2.3 Molecular Construction of Newly Discovered Sensory Receptors

Amazing acknowledgements about vision and olfaction have progressed in those 50 years, but concern about the other senses, even the existence of receptors, cannot be identified. However, recent progress of molecular biology has made it clear that the receptors concerned about olfaction, gustation, and nociperception are exactly existing.

About the olfaction, there are several hundred types of olfactory receptors in order to distinguish different several hundred thousand smells. Groups of olfactory receptors were discovered in 1991, and this has accelerated our understanding about the distinction of olfactory molecules in the brain, such as "olfactory receptor map" in the olfactory bulb.

Human feels a sensation of pain by accepting nociceptive stimuli, and one of those nociceptors is the capsaicin receptor, causing pain with hot taste. On the other hand, many kinds of nociceptors such as P2X3 receptor or proton-sensitive ion channel-type receptor were also cloned, and the investigation about thermoesthesia and algesthesia is now conducted. Among these receptors, we have investigated about capsaicin receptor, one of the nociceptors and taste receptors in the larynx.

1.1.3 Nociceptors

Algetic stimuli are input to the brain stem through primary afferent sensory nerve and end up to be identified as pain in cerebral cortex. Nerve fibers in the primary afferent sensory nerves that participate in algesthesia are unmyelinated C-fibers and myelinated Aδ-fibers. These C-fibers or Aδ-fibers originated from small- or middle-sized cells in spinal ganglia, and there are mainly polymodal receptors or high liminal mechanoreceptors in the terminal of their neural terminals. On the other hand, thigmesthesia and baresthesia that don't cause pain are transmitted by myelinated Aβ-fibers. It is said that Aβ-fibers may participate in allodynia. Fine primary afferent sensory nerves (C-fibers or Aδ-fibers) function only in acceptance and transmission of pain.

Stimuli that elicit nociperception to the body include mechanical stimuli, thermal stimuli, and chemical stimuli. As the receiver of these stimuli, nociceptors such as capsaicin receptor, ATP receptor, acid-sensing ion channel receptor, and so on were cloned, and the studies about nociceptors are now making rapid progress.

1.1.3.1 Capsaicin Receptor

Capsaicin, the main ingredient in chili peppers, elicits pain with a spicy taste (Fig. 1.1). A functional cDNA encoding a capsaicin receptor has isolated from sensory neurons with an expression cloning strategy based on calcium influx [1]. Because capsaicin has the vanillyl base as its

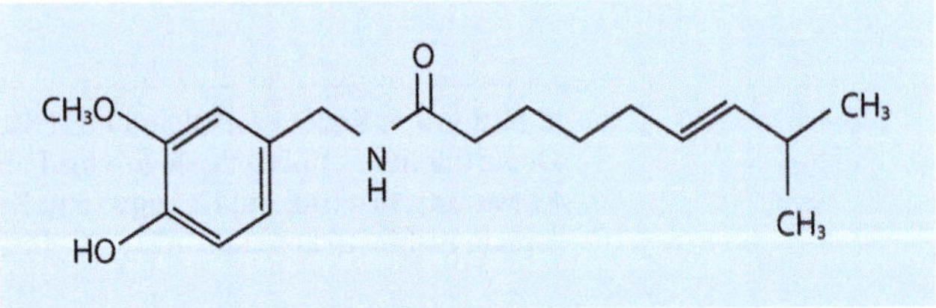

Fig. 1.1 Molecular formula of capsaicin

structure, the capsaicin receptor was named vanilloid receptor subtype 1 (VR1) previously and now called transient receptor potential channel, vanilloid subfamily 1 (TRPV1) (Fig. 1.2). The cloned vanilloid receptor is also activated by heat of 43 °C or more. Analysis of heat-evoked single-channel currents in excised membrane patches showed that heat gates TRPV1 directly. Moreover, protons decrease the temperature threshold for TRPV1 activation, such that even moderately acidic conditions (PH<5.9) activate TRPV1 at room temperature. Thus TRPV1 has been known that it is a nonselective cation channel and be activated not only by capsaicin but also by noxious heat and protons, and it has been suggested that it is a polymodal nociceptor (Fig. 1.3).

TRPV1 transcripts are located in small to medium diameter primary sensory neurons (C-fiber neurons) in dorsal root and trigeminal ganglia [2]. Intense immunoreactivity was observed in the terminals of afferent fibers projecting to superficial layers of the spinal cord dorsal horn and the trigeminal nucleus caudalis. In the spinal cord, the densest staining was found in laminae 1 and 2.

The gene of a TRPV1 homologue was also cloned and named vanilloid receptor-like protein 1 (VRL1) [3]. It is now called TRPV2, one of the subtypes of transient receptor potential cation channels. However, TRPV2 does not have sensitivity to capsaicin or protons; it can be activated only by heat exceeding 52 °C (Fig. 1.3).

1.1.4 Taste Receptors

Gustation provides information about the foods and liquids. In order to avoid intakes of harmful materials and take in necessary nourishments, we have acquired the diversity of

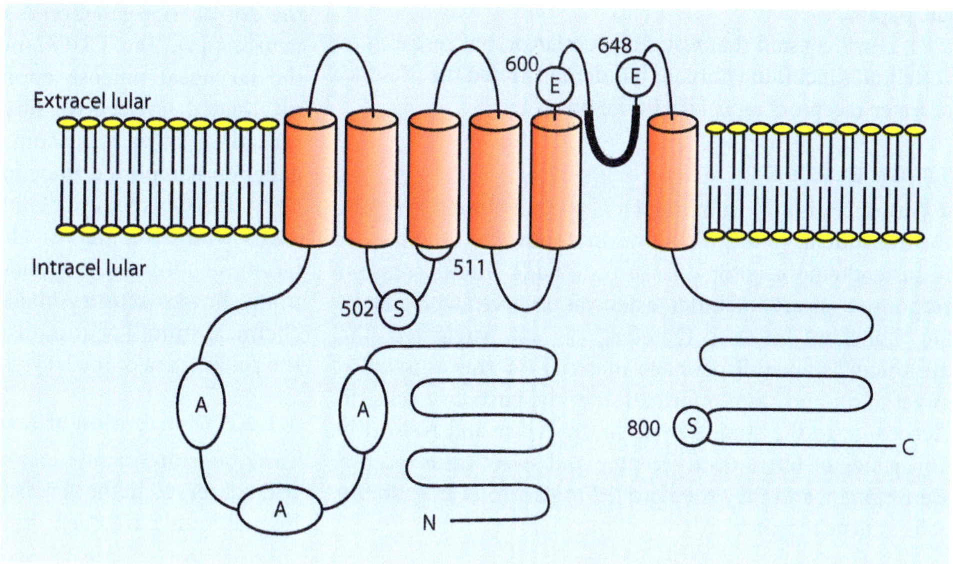

Fig. 1.2 Membrane topology model of TRPV1

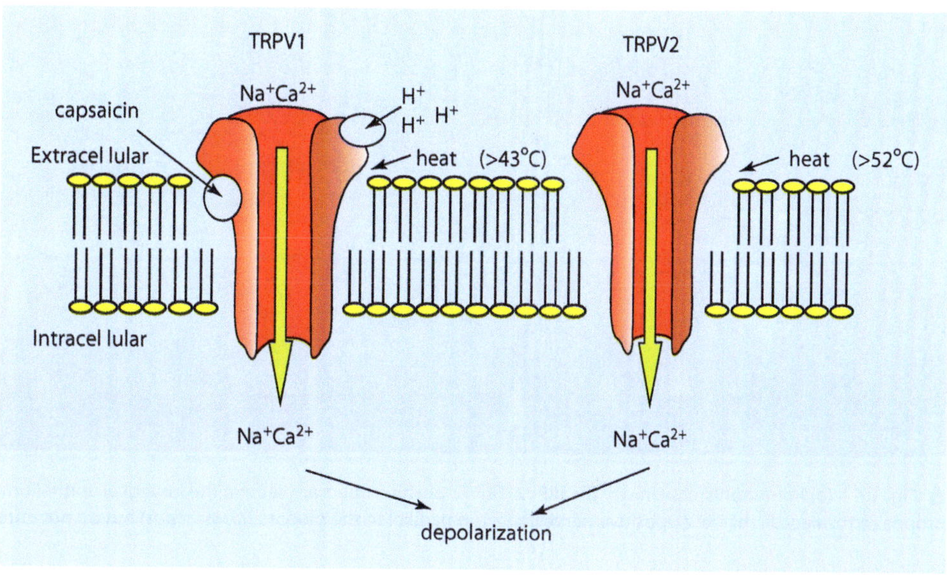

Fig. 1.3 Activation of TRPV1 and TRPV2

gustatory acceptance in the process of evolution. It is known that taste buds which detect and discriminate various chemical substances work as sensory receptors. They are mainly distributed over the superior surface of the tongue, but have also been observed in the palate, pharynx, and larynx in many species of mammals. Gustatory stimuli are sensed by gustatory cells. Dissolved chemicals contacting the microvilli at taste pore bind to receptor proteins which distribute peculiarly in gustatory cells, and the receptor cells release neurotransmitter. The gustatory information is transmitted to the appropriate portions of the primary sensory cortex through four neurons.

We mammals recognize the four primary taste sensations: salt, sour, sweet, and bitter. In addition to that, humans can detect two additional tastes: umami and water [4].

Umami is produced by receptors sensitive to the presence of amino acids, especially glutamate, small peptides, and nucleotides. The distribution of these receptors is not known in detail, but they are present in taste buds of the circumvallate papillae.

It is widely said that water has no flavor, but research on vertebrates including humans has demonstrated the presence of water receptors especially in the pharynx.

1.1.4.1 Gustducin

α-Gustducin is an α-subunit of a G protein closely related to the transducins [5–8], which was first shown to be expressed in taste chemoreceptor cells of taste buds. The decrease of response to sweet and bitter tastes was observed with gustducin knockout mouse; therefore, it was suggested that gustducin is an indispensable material for transmission of sweet and bitter taste information at the gustatory cells [6]. According to the studies by Chandrashekar and Nelson [9, 10], genes of bitter taste receptor and sweet taste receptor were cloned, and they were proved to play roles in gustatory cells as taste receptors.

1.1.5 Distribution of Receptors

1.1.5.1 Distribution of Capsaicin Receptor in the Larynx

In order to understand the function of vanilloid receptors in the laryngeal innervation, we examined the distribution of capsaicin receptors in the rat larynx using TRPV1 and TRPV2 immunohistochemistry [11]. TRPV2-positive nerve fibers were detected in the laryngeal mucosal epithelium and along the lamina propria (◘ Fig. 1.4a). TRPV1-positive nerve fibers were seen in the lamina propria running parallel to the mucosa but do not enter the epithelium (◘ Fig. 1.4b). Both TRPV1- and TRPV2-positive neurons were also found in the intralaryngeal ganglia.

Only TRPV2-positive nerve fibers were seen to enter the laryngeal mucosa epithelium and could be considered free nerve endings. In mucosal epithelium most neurons are unmyelinated, but some myelinated fibers are proved to exist. It has been reported that TRPV2 immunoreactivity in the dorsal root ganglion is seen in Aδ myelinated sensory neurons [1]. The TRPV2-positive nerve fibers detected in the laryngeal mucosa epithelium may be comparable to myelinated nociceptive fibers that rapidly transmit sharp sensations of pain. In contrast, TRPV1 in the dorsal root ganglion is reported to be localized in small- and medium-sized neurons that are mainly C-fiber sensory neurons. The TRPV1-positive nerve fibers running parallel to the laryngeal mucosa epithelium may be similar slow unmyelinated sensory fibers that convey information from chemical stimuli or stimuli associated with inflammation in the submucosa of the larynx.

1.1.5.2 Distribution of Taste Receptors in the Larynx

Taste buds are mainly located on the tongue, but have also been observed in the larynx [12–19]. Although laryngeal and lingual buds have been described as morphologically quite

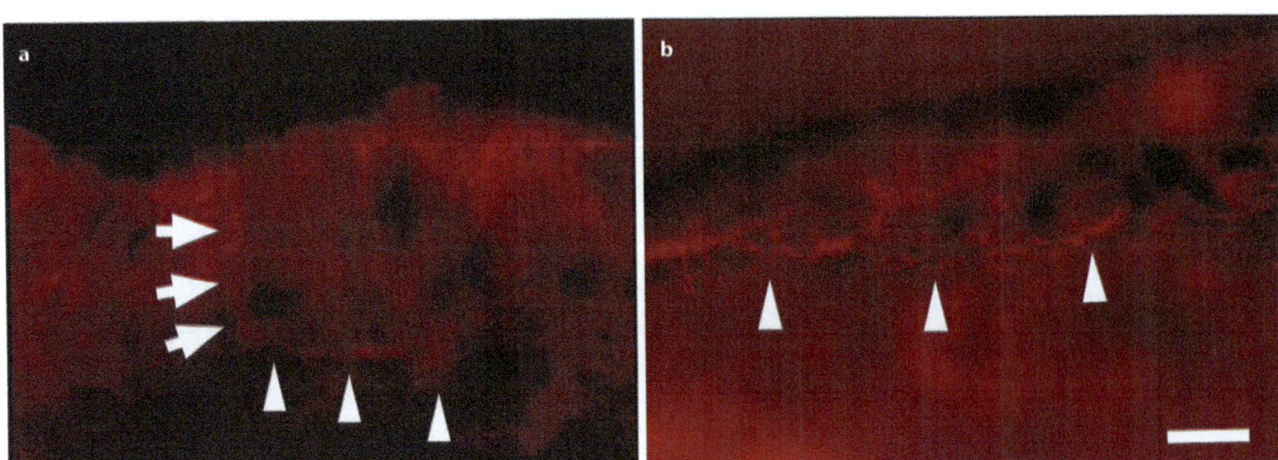

◘ **Fig. 1.4** (**a**) Immunohistochemistry for TRPV2. TRPV2-positive fibers are seen in the laryngeal epithelium (*arrows*) and along the lamina propria (*arrowheads*). (**b**) TRPV1-positive nerve fibers run parallel to the mucosa (*arrowheads*) but do not enter the epithelium

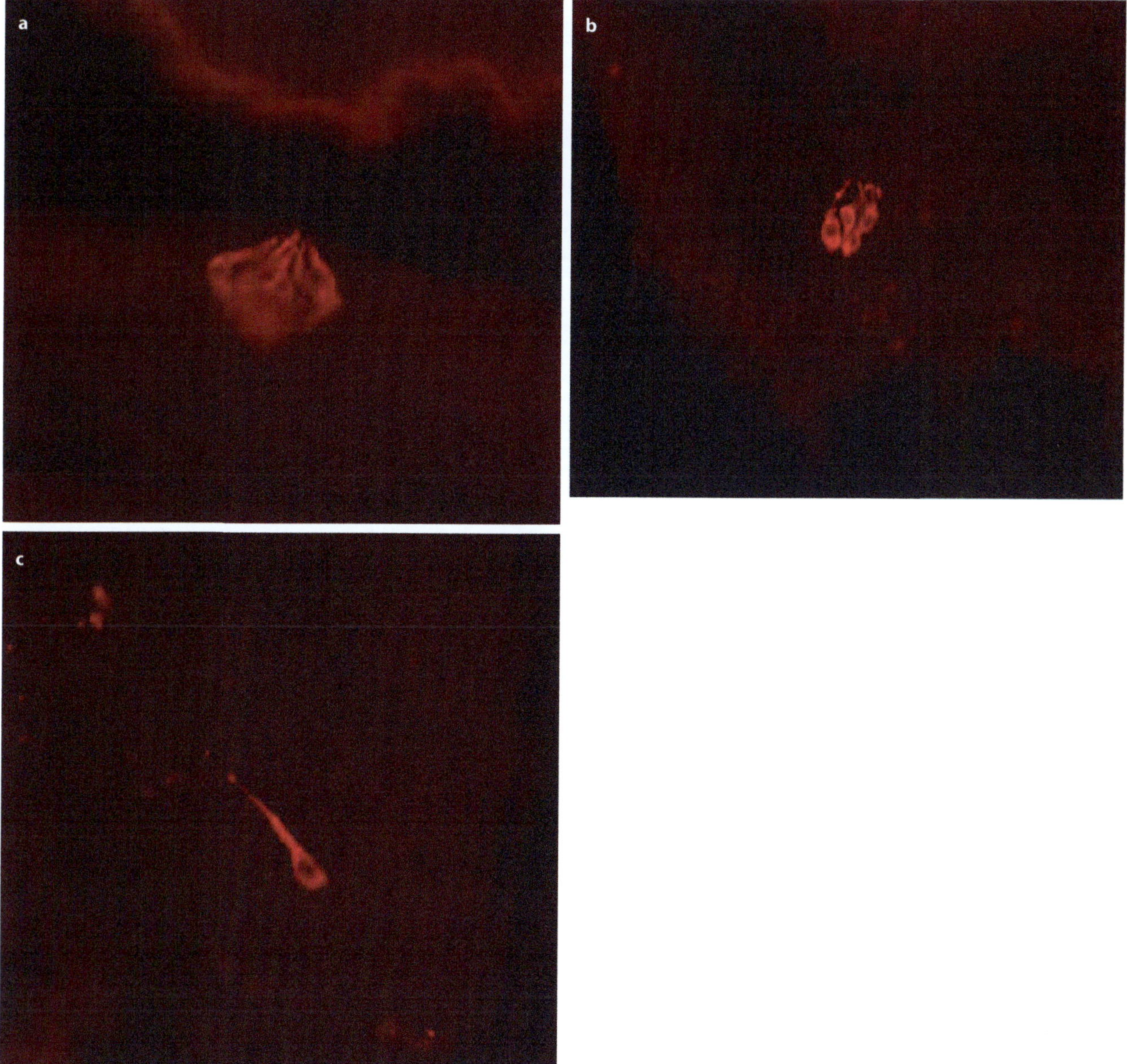

Fig. 1.5 (**a**) α-Gustducin-immunoreactive cells are congregated, presenting typical bud form. (**b**) Immunohistochemistry for α-gustducin of chemosensory clusters. (**c**) Immunohistochemistry for α-gustducin of solitary chemosensory cells

similar structures, functional differences may exist between them. We investigated the distribution of α-gustducin-immunoreactive taste cells and its age-related changes in the murine larynx [20]. Three different morphologic types of α-gustducin-immunoreactive structures were seen [21, 22]: typical gemmal forms (Fig. 1.5a), clusters composed of two or three cells (chemosensory clusters, CCs) (Fig. 1.5b), and isolated immunoreactive cells (solitary chemosensory cells, SCCs) (Fig. 1.5c). There were about 80 α-gustducin-immunoreactive structures in average per larynx, which were located mainly on the laryngeal surface of the epiglottis and on the arytenoids. They were distributed most densely close to the caudal base of the laryngeal surface of the epiglottis, extending along the aryepiglottic folds and arytenoids. The total number of these α-gustducin-immunoreactive structures did not show any age-related changes, while the percentage of SCCs in 5-week-old rats was significantly larger than the respective number in 8-, 14-, and 21-week-old rats. According to a previous study in the tongue of the newborn rat, the presence of SCCs is accompanied by a rapid development of intrinsic neurons [23]. It was suggested that different pathways (i.e., gustatory and solitary chemosensory cell system) are involved in the oral chemoreception and that a primitive system of SCCs may develop and be replaced by

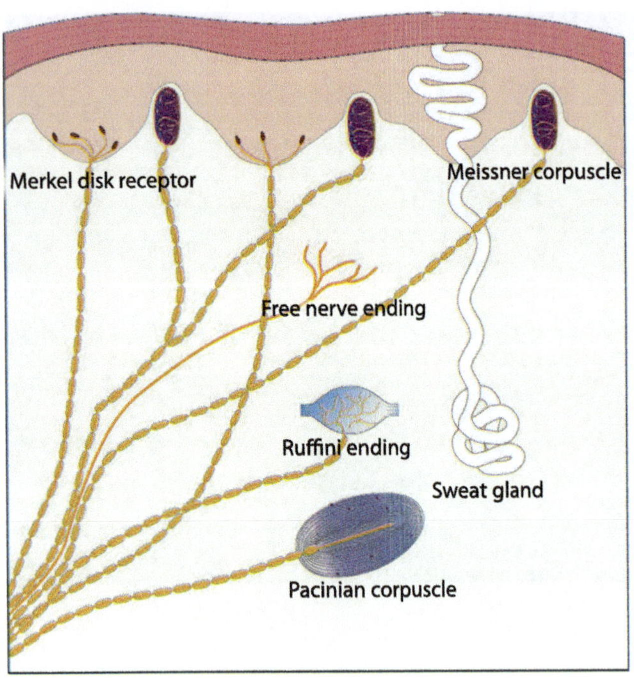

Fig. 1.6 Tactile receptors in the skin

taste buds [23]. It is conceivable that laryngeal SCCs also may be replaced to CCs or buds during development. It is suggested that the SCCs found facing the laryngeal airway at the base of the epiglottis may be specialized chemosensors protecting the airway, which is an essential function for survival of the animal, as compared to the gustatory function of their lingual counterparts, while laryngeal CCs or buds are highly differentiated chemosensory cells and may be potentially able to participate in taste perception in the larynx.

1.2 Sensory Nerve Endings (Fig. 1.6)

1.2.1 Introduction

Sensory receptors [24, 25] are specialized cells or cell processes that provide the central nervous system with information about internal or external conditions of the body. A sensory receptor detects an arriving stimulus and translates it into an action potential that can be transmitted to the CNS.

The widely distributed general sensory receptors are involved in temperature, pressure, touch, vibration, position sense, and pain. Anatomically, these receptors are either free dendritic endings or encapsulated dendritic endings.

1.2.2 Types of Sensory Nerve Endings

1.2.2.1 Free Nerve Endings

The simplest receptors are the dendrites of sensory neurons, and the branching tips of them, called free nerve endings, are not protected by accessory structures. Most of these fibers are unmyelinated. They are particularly abundant in epithelia and connective tissue, but found in nearly everywhere in the body, dermis, subcutaneous tissue, fascia, periosteum, serous tunics, tunica adventitia, choroidea, and the like. They are known to work not only as thermoreceptors or mechanoreceptors that accept temperature (heat and cold) or touching sense but also polymodal nociceptor toward noxious heat or mechanical pressure which caused severe damage in tissues.

1.2.2.2 Ruffini's Corpuscles

Ruffini's corpuscles, which are located in the dermis, subcutaneous tissue, and joint capsules, contain a spray of dendritic endings enclosed by a flattened capsule. In the capsule, a network of dendrites is intertwined with the collagen fibers. According to the unique form, they are also called Ruffini's cylinder or spindle. They are sensitive to pressure and distortion of the skin. As they are located in joint capsules and the surrounding connective tissue, they probably play a role to respond to deep and continuous pressure. The other previous study referred to the potential as thermoreceptors because they were widely distributed in subcutaneous tissue in the body. They are now regarded to work as slow-adapting mechanoreceptors and respond to the definite pressure toward the dermis.

1.2.2.3 Merkel Discs

In 1875, Merkel discovered large-scale clear cells scattered in the epidermal tissue of pig nose and named them "tactile cell" according to their coincident distribution with the sensitive tactual part. The functional role of them has not been investigated completely, and they are now called Merkel cells. Certain free dendritic processes of a single myelinated afferent fiber make close contact with enlarged, disc-shaped epidermal cells (Merkel cells) to form Merkel discs. Merkel discs are fine touch and pressure receptors, which lie in the deeper layers of the skin epidermis. They are extremely sensitive tonic receptors with very small receptive fields. Iggo classified Merkel discs as type 1 SA unit and Ruffini's corpuscles as type 2 SA unit, but opinions are divided on the matter to classify them into two types.

1.2.2.4 Pacinian Corpuscles

Pacinian corpuscles, also called large lamellated corpuscles, are the largest of the corpuscular receptors which are found in mammals. In 1741 they were discovered in the fingers with the naked eyes by Vater. The entire corpuscle may reach 4 mm in length and 2 mm in diameter, and some are visible to the naked eyes as white, egg-shaped bodies. A single dendrite is surrounded by up to 60 concentric layers of collagen fibers and supporting cells, which in turn are enclosed by a connective tissue capsule. They are fast-adapting receptors and most sensitive to high-frequency vibration. They adapt quickly because distortion of the capsule soon relieves pressure on the sensory process. They are scattered deep in the dermis, notably in the fingers, mammary glands, and external genitalia. They are also found in the superficial and deep fascia, joint capsules, mesenteries, pancreas, and the walls of the urethra and urinary bladder.

1.2.2.5 Meissner's Corpuscles

Meissner's corpuscles were discovered by Wagner and Meissner in the human skin in 1852. They are found just beneath the skin epidermis in the dermal papillae. They involve a few spiraling dendrites surrounded by a thin egg-shaped capsule of connective tissue, measuring roughly 30–100 μm in length. They are most abundant in sensitive and hairless skin areas such as fingertips, palms, soles of the feet, and nipples. They are also found in eyelids, lips, tip of tongue, and pharyngeal mucosa. They are also called tactile corpuscles and perceive sensations of fine touch and pressure. These corpuscles are revealed to become bigger and bigger and show irregular shape with aging, but 80 % of them disappeared. According to this fact, it is considered that Meissner's corpuscles are formed and grow in proportion to living environment and disappear in due time.

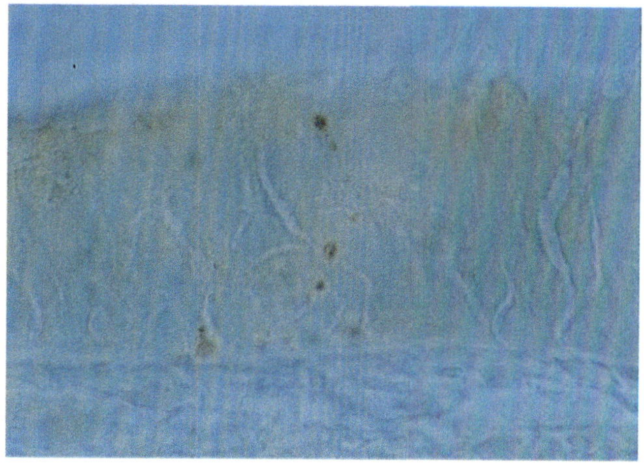

Fig. 1.7 Substance P-immunoreactive nerve fibers in the canine larynx

1.2.3 Sensory Nerve Endings in the Larynx

The glottis is the last barrier to protect the airway from errant swallowing, and the sensory nerve endings that distribute in the laryngeal mucosa are significantly associated with the initiation of the laryngeal reflex reaction to protect the airway. It is known that the distribution density of sensory units in the larynx is not uniform and the sensory nerve endings are especially abundant along the base of epiglottis and on arytenoids [26, 27]. At the glottis mechanoreceptors are localized in the front and back of the vocal folds, and it is suggested that they participate not only with the laryngeal protection reflex but also with the regulation of laryngeal muscles in breath or phonation [28].

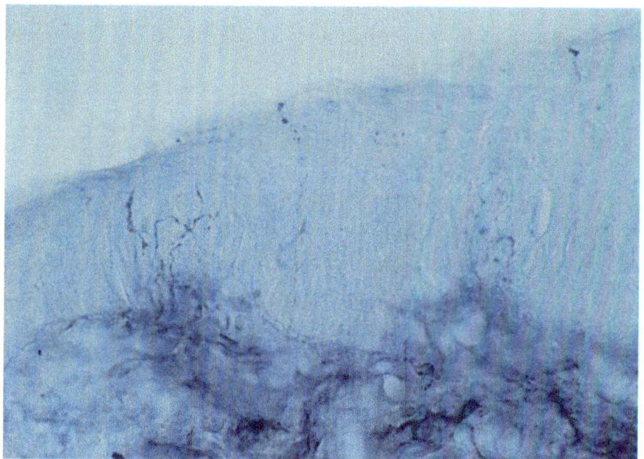

Fig. 1.8 CGRP-immunoreactive nerve fibers in the canine larynx

1.2.4 Non-noradrenergic, Non-cholinergic Transmitters in the Laryngeal Sensory Nerve Endings

1.2.4.1 Neuropeptide in the Laryngeal Sensory Nerve Endings

We reported that the superior laryngeal nerve and the inferior laryngeal nerve contained substance P (SP)-immunoreactive nerve fibers in the canine larynx, and SP might be involved in the laryngeal sensory innervation system through these two nerves (Fig. 1.7) [29]. Many SP-immunoreactive nerve fibers were observed within the epithelial layer and in the connective tissue below the epithelium of the laryngeal mucosa, and there were no differences in density of the distribution between the ventricle, glottis, and subglottis [30].

We also made it clear that calcitonin gene-related peptide (CGRP) plays an extremely significant role in the canine laryngeal sensory innervation [31]. CGRP-immunoreactive nerve fibers were found in every region of the larynx, especially in the epiglottis, arytenoids and subglottis (Fig. 1.8). CGRP-immunoreactive nerve fibers were found more frequently than SP-immunoreactive nerve fibers, and these findings suggest that CGRP plays a more important role. Many nerve fibers with varicosities appeared to terminate in epithelial layer of laryngeal mucosa and form free nerve endings. In the epiglottis and arytenoids, taste buds were observed and they were densely innervated by the CGRP-immunoreactive nerve fibers. Abundant CGRP-immunoreactive nerve fibers were observed in this region: the epiglottis and arytenoid, and this suggests that they are concerned in mechanoreception and chemoreception. It is generally thought that laryngeal taste buds which are located in arytenoids and the epiglottis may work as chemosensory detectors to initiate the reflex reaction to protect the airway

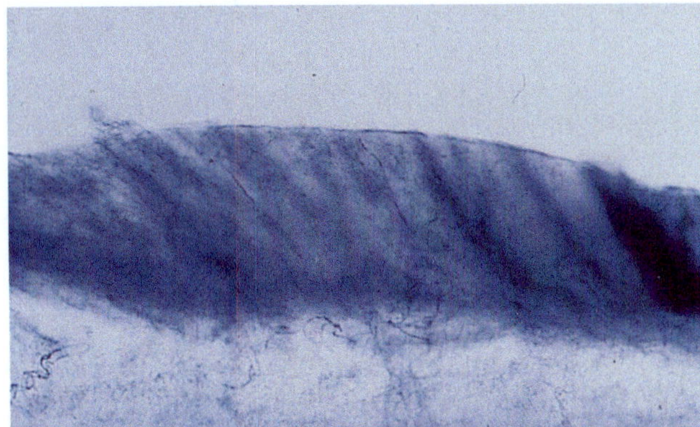

Fig. 1.9 NADPHd-immunoreactive nerve fibers in the murine larynx

from mis-swallowing [14], and the dense innervation of taste buds by the CGRP-immunoreactive nerve fibers strongly suggests the importance of CGRP in the laryngeal sensory nervous system.

1.2.4.2 Nitric Oxide (NO) in the Laryngeal Sensory Nerve Endings

We evaluated the involvement of nitric oxide (NO) in the murine laryngeal innervation with NADPH-diaphorase (NADPHd) histochemistry [32]. We found NADPHd-positive nerve fibers in every region of the larynx, especially abundant in the lamina propria (Fig. 1.9). There, some of these fibers were associated with blood vessels or laryngeal glands, suggesting that they might modulate blood flow and exocrine secretion of the larynx. A small number of NADPHd-positive nerve fibers were detected in the epithelia of the mucosa, appearing to terminate to the mucosal surface with varicosities. NO may take part in nociperception of the larynx.

References

1. Caterina MJ, Schumacher MA, Tominaga M, Rosen TA, Levine JD, Julius D. The capsaicin receptor: a heat-activated ion channel in the pain pathway. Nature. 1997;389:816–24.
2. Tominaga M, Caterina MJ, Malmberg AB, Rosen TA, Gilbert H, Skinner K, Raumann BE, Basbaum AI, Julius D. The cloned capsaicin receptor integrates multiple pain-producing stimuli. Neuron. 1998;21:531–43.
3. Caterina MJ, Rosen TA, Tominaga M, Brake AJ, Julius D. A capsaicin receptor homologue with a high threshold for noxious heat. Nature. 1999;398:436–41.
4. Lindemann B. Receptors and transduction in taste. Nature. 2001; 413:219–25.
5. McLaughlin SK, McKinnon PJ, Margolskee RF. Gustducin is a taste-cell-specific G protein closely related to the transducins. Nature. 1992;357:563–9.
6. Wong GT, Gannon KS, Margolskee RF. Transduction of bitter and sweet taste by gustducin. Nature. 1996;381:796–1003.
7. Boughter JD, Pumplin DW, Yu C, Christy RC, Smith DV. Differential expression of α-gustducin in taste bud populations of the rat and hamster. J Neurosci. 1997;17:2852–8.
8. Yang R, Tabata S, Crowley HH, Margolskee RF, Kinnamon JC. Ultrastructural localization of gustducin immunoreactivity in microvilli of type II taste cells in the rat. J Comp Neurol. 2000;425:139–51.
9. Chandrashekar J, Mueller KL, Hoon MA, Adler E, Feng L, Guo W, Zuker CS, Ryba NJ. T2Rs function as bitter taste receptors. Cell. 2000;100:703–11.
10. Nelson G, Hoon MA, Chandrashekar J, Zhang Y, Ryba NJ, Zuker CS. Mammalian sweet taste receptors. Cell. 2001;106:381–90.
11. Koike S, Uno T, Bamba H, Shibata T, Okano H, Hisa Y. Distribution of vanilloid receptors in the laryngeal innervation. Acta Otolaryngol. 2004;124:515–9.
12. Andrew BL, Oliver J. The epiglottal taste buds of the rat. J Physiol. 1951;114:48–9.
13. Khaisman EB. Particular features of the innervation of taste buds of the epiglottis in monkeys. Acta Anat. 1976;95:101–15.
14. Bradley RM, Cheal ML, Kim YH. Quantitative analysis of developing epiglottal taste buds in sheep. J Anat. 1980;130:25–32.
15. Palmieri G, Asole A, Panu R, Sanna L, Farina V. On the presence, structure and probable functional role of taste buds located on the laryngeal surface of the epiglottis in some domestic animals. Arch Anat Histol Embryol. 1983;66:55–66.
16. Miller IJ, Smith DV. Quantitative taste bud distribution in the hamster. Physiol Behav. 1984;32:275–85.
17. Shin T, Nahm I, Maeyama T, Miyazaki J, Matsuo H, Yu Y. Morphologic study of the laryngeal taste buds in the cat. Laryngoscope. 1995;105: 1315–21.
18. Shrestha R, Hayakawa T, Das G, Thapa TP, Tsukamoto Y. Distribution of taste buds on the epiglottis of the rat and house shrew, with special reference to air and food pathways. Okajimas Folia Anat Jpn. 1995;72:137–48.
19. Yamamoto Y, Atoji Y, Hobo S, Yoshihara T, Suzuki Y. Morphology of the nerve endings in laryngeal mucosa of the horse. Equine Vet. 2001;33:150–8.
20. Nishio T, Koike S, Okano H, Bamba H, Hisa Y. Age-related expression of α-gustducin in the rat larynx. Ann Otol Rhinol Laryngol. 2006;115:387–93.
21. Sbarbati A, Merigo F, Benati D, Tizzano M, Bernardi P, Osculati F. Laryngeal chemosensory clusters. Chem Senses. 2004;29:683–92.
22. Sbarbati A, Merigo F, Benati D, Tizzano M, Bernardi P, Crescimanno C, Osculati F. Identification and characterization of a specific sensory epithelium in the rat larynx. J Comp Neurol. 2004;475:188–201.
23. Sbarbati A, Crescimmano C, Bernardi P, Benati D, Merigo F, Osculati F. Postnatal development of the intrinsic nervous system in the circumvallate papilla-von Ebner gland complex. Histochem J. 2000;32:483–8.
24. Martini FH. Fundamentals of anatomy and physiology. 5th ed. Upper Saddle River: Prentice-Hall; 2001. p. 527–37.
25. Marieb EN. Human anatomy and physiology. 5th ed. San Francisco: Addison Wesley Longman; 2001. p. 475–81.
26. Orey AT. Functional analysis of sensory units innervating epiglottis and larynx. Exp Neurol. 1968;20:366–83.
27. Davis PJ, Nail BS. Quantitative analysis of laryngeal mechanosensitivity in the cat and rabbit. J Physiol. 1987;388:467–85.
28. Wyke BD, Kirchner JA. Neurology of the larynx. In: Hinchcliffe R, Harrison D, editors. Scientific foundation of otolaryngology. London: Heinenmann; 1976. p. 546–74.
29. Hisa Y, Sato F, Fukui K, Ibata Y, Mizukoshi O. Substance P nerve fibers in the canine larynx by PAP immunohistochemistry. Acta Otolaryngol. 1985;100:128–33.
30. Hisa Y, Toyoda K, Uno T, Murakami Y, Ibata Y. Localization of the sensory neurons in the canine nodose ganglion sending fibers into the internal branch of the superior laryngeal nerve. Eur Arch Otorhinolaryngol. 1991;248:265–7.
31. Hisa Y, Uno T, Tadaki N, Murakami Y, Okamura H, Ibata Y. Distribution of calcitonin gene-related peptide nerve fibers in the canine larynx. Eur Arch Otorhinolaryngol. 1992;249:52–5.
32. Hisa Y, Tadaki N, Uno T, Koike S, Tanaka M, Okamura H, Ibata Y. Nitrergic innervation of the rat larynx measured by nitric oxide synthase immunohistochemistry and NADPH-diaphorase histochemistry. Ann Otol Rhinol Laryngol. 1996;105:550–4.

Muscle Spindles and Intramuscular Ganglia

Shinobu Koike, Shigeyuki Mukudai, and Yasuo Hisa

S. Koike • S. Mukudai • Y. Hisa (✉)
Department of Otolaryngology-Head and Neck Surgery, Kyoto Prefectural University of Medicine, Kawaramachi-Hirokoji, Kamigyo-ku, Kyoto 602-8566, Japan
e-mail: largo@koto.kpu-m.ac.jp; yhisa@koto.kpu-m.ac.jp

Y. Hisa (ed.), *Neuroanatomy and Neurophysiology of the Larynx*, DOI 10.1007/978-4-431-55750-0_2

2.1 Muscle Spindles

Muscle spindles, which are known to be sensory receptors of stretch, exist in the connective tissue between the muscle fibers of skeletal muscles. They are part of the reflex arch controlling muscle contraction. Studies incorporating such techniques as light microscopy, electron microscopy, and histochemistry have shown that the intrafusal muscle fibers found inside the muscle spindle capsule consist of four types, each with a characteristic innervation. In the human larynx, muscle spindles are seen in the interarytenoid and posterior cricoarytenoid muscles, although the sizes of the muscle spindles are smaller than those seen in other skeletal striated muscles. Much fewer muscle spindles are seen in the intrinsic laryngeal muscles compared with other skeletal muscles, and much of their function is unknown.

2.1.1 Introduction

Interest was first focused on muscle spindles in the 1880s, when Kerschner [1], Ruffini [2], and Sherrington [3] reported studies of small bundles of muscle fibers in the connective tissue between muscle fibers, differing from the usual striated muscle fibers outside the spindle-shaped capsules, and with possible sensory function. The existence of muscle spindles had been known since they were morphologically pointed out by Koelliker in 1862, but were at first considered centers of muscle growth. It is now known that the muscle spindle is a sensory receptor end organ, similar to the Golgi tendon organ, Pacinian corpuscle, and free nerve endings, and provides sensory information on muscle tension.

The existence of muscle spindles in the human intrinsic laryngeal muscles was reported by Goerttler [4] and Paulsen [5] in the 1950s. Researchers such as Baken [6] and Grim [7] followed with wide histological investigation and showed that muscle spindles existed in all intrinsic laryngeal muscles. In 1987, Katto [8] and Hirayama [9] reported electron microscopic studies on the fine structures of muscle spindles.

2.1.2 Muscle Spindles in General

2.1.2.1 Distribution of Muscle Spindles

Muscle spindles are seen in all skeletal muscles, more abundantly in muscles requiring sustained contraction such as cervical muscles and in muscles related to fine movement such as the lumbrical muscles in the hands and feet and muscles related to eye movement. Cooper [10] compared the density of muscle spindle distribution between various muscles and reported that 15 or more spindles per gram muscle were found in spindle abundant muscles such as the lumbrical muscles (12.2–19.7 spindles/g). Muscles related to coarse movement had a sparse distribution below seven spindles per gram muscle such as 1.4 spindles/g in the latissimus dorsi muscle. Muscle spindles are known to be more abundant in the muscle tissue close to the tendons, but are also seen in the muscle belly and even in the intermuscular septum and fascia. The arrangement of muscle spindles has been classified into three types: solitary single type spindles, tandem types with multiple spindles in series, and compound types with several parallel spindles more or less fused together [11].

2.1.2.2 General Morphology of Muscle Spindles

Muscle spindles are found in the connective tissue between the muscle fibers of skeletal muscles. They are, as their name implies, spindle shaped with a length of about 2–20 mm and a width of about 50–200 μm. A muscle spindle is composed of several striated muscle fibers (intrafusal muscle fibers) innervated by nerve fibers and is surrounded by a strong capsule proper of connective tissue. A circumferential lymph space (periaxial lymph space) exists between the capsule and its contents, wide at the equator of the spindle and narrow near the ends or poles. Nuclei are collected near the equator of the spindle. The capsule proper is composed of collagen fibers and fibroblasts and is continuous with the perimysium at the poles of the spindle and with the perineurium at places where a nerve enters the spindle. A thin layer of delicate connective tissue fibers directly covers the intrafusal muscle fibers. A gelatinous fluid rich in glycosaminoglycans fills the space between this inner capsule (axial sheath) and the capsule proper. The fine intrafusal muscle fibers differ from the heavy extrafusal muscle fibers outside the spindles.

Two types of intrafusal muscle fibers have been identified in mammalian muscle tissue by morphological studies. Nuclear bag type fibers are approximately 20 μm wide and 7–8 mm long, with many small nuclei collected near the equator of the spindle. This together with the lack near the equator of stripes normally seen in striated muscles gives them a bulging bag-like appearance. Nuclear chain type fibers are smaller in diameter and shorter, typically about 10 μm wide and 4 mm long, and do not have the equatorial bulge. Nuclear chain type fibers are positioned close to the axis of the spindle, while nuclear bag type fibers are situated closer to the capsule proper. Each muscle spindle usually contains 1–4 nuclear bag type muscle fibers and 1–10 nuclear chain type muscle fibers. Motor innervations by γ motor axons are known to exist in both nuclear bag type and nuclear chain type muscle fibers [12, 13].

Based on ATPase reactivity [14–19] and oxidative enzymatic activity [20, 21], the two types of intrafusal muscle fibers in human muscle tissue have been further distinguished into three types. However, the two nuclear bag fiber types in each report may not represent the same two subgroups of nuclear bag type fibers since the histochemistry employed differs in some of the studies. Some of the confusion may have been caused by the fact that different regions (polar, equatorial, or intermediate) of the intrafusal muscle fibers have different histochemical properties. Based on studies on mammalian muscle, Banks [22] reviewed and summarized the classifications and concluded that the nuclear

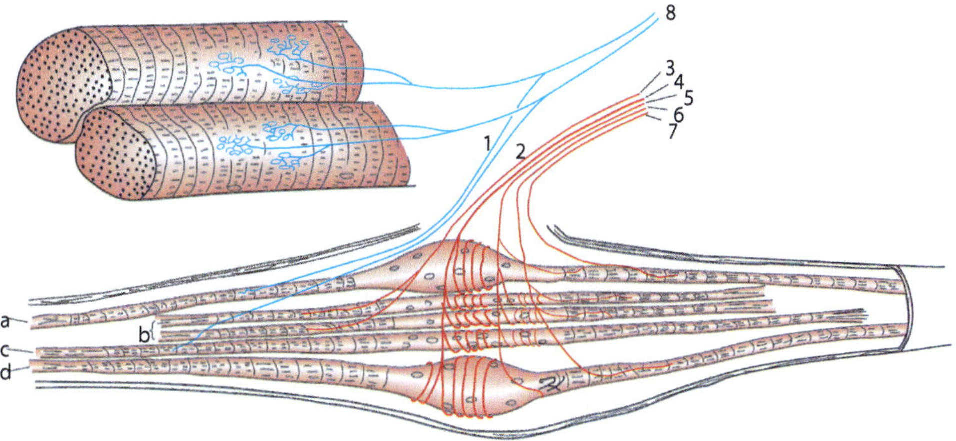

Fig. 2.1 (**a**) nuclear bag$_1$ fiber, (b) short-chain fiber, (c) long-chain fiber, (d) nuclear bag$_2$ fiber, 1 dynamic β motor neuron (P_2 end plate), 2 static β motor neuron (P_1 end plate), 3 static γ_2 motor neuron (trail ending), 4 type Ia sensory fiber (primary annulospiral ending), 5 type IIa sensory fiber (flower-spray ending), 6 static γ_2 motor neuron, 7 dynamic γ_1 motor neuron (P_2 end plate), 8 α motor neuron (Revised from [38])

bag type fibers were found to consist of two different types of muscle fibers, the nuclear bag$_1$ and bag$_2$ muscle fibers. The classification was mainly based on results of myofibrillar ATPase following alkaline preincubation [23]. Nuclear bag$_2$ fibers have a moderate to high alkaline ATPase reactivity and very clear myofibril M-lines are evident near the poles. The nuclear bag$_1$ muscle fibers in comparison are thinner, shorter, and rich in mitochondria and oxidizing enzymes and have fewer sarcoplasmic reticula and low ATPase reactivity. The nuclear bag$_1$ fibers lack the M-lines in their myofibrils in their equatorial and intracapsular polar regions. Nuclear chain type fibers have a high ATPase reactivity, but have few mitochondria and low oxidizing enzyme reactivity. Sarcoplasmic reticula are well developed, and the myofibril M-lines are clearly evident in the nuclear chain type fibers. The fiber composition of the axial sheath is known to differ between the nuclear bag$_1$ and bag$_2$ muscle fibers. Nuclear bag$_2$ muscle fibers contain many elastic fibers in the axial sheath, while nuclear bag$_1$ muscle fibers do not. Saito [20] has suggested the existence of a minor subtype of nuclear chain type muscle fiber in humans only.

2.1.2.3 Innervation of Muscle Spindles

There are two types of afferent nerve fibers innervating the intrafusal muscle fibers. Group Ia axons conduct sensory inputs from the primary annulospiral sensory nerve endings encircling the intrafusal muscle fibers at the equator. Group II axons are secondary sensory fibers from nerve endings between the equator and either pole. Both group Ia fibers and group II fibers are myelinated fibers and do not lose their myelin sheath until penetrating deeply into the muscle spindle, beyond the periaxial lymph space. Each group Ia axon branches and innervates multiple intrafusal muscle fibers. The cytoplasm is abundant in mitochondria, and synaptic-like vesicles that may be part of feedback pathways acting as gain control mechanisms are seen in the sensory endings [24]. Group II secondary sensory nerve fibers branch out and end in "flower-spray endings" adjacent to the equatorial region on either pole side, mostly on nuclear chain type muscle fibers [25–27].

Three types of efferent motor innervation of the intrafusal muscle fibers are known, each with its characteristic nerve ending (Fig. 2.1). Efferent γ axons from γ motor neurons are distributed on the intrafusal muscle fibers, terminating adjacent to the equator. The nerve fibers ending in trail endings near the equator are the γ_2 fibers. The γ_1 nerve fibers end in plate P_2 endings between the equatorial and polar regions. In the 1970s, a third origin of efferent innervation was discovered. Motor β fibers originating in branches from α motor fibers innervating the extrafusal muscle fibers were found to end in plate endings close to the poles of the muscle spindle [28, 29]. These plate endings are called plate P_1 endings. Variation over species and muscle exists, so not all muscle spindles share this combination of efferent innervation. Multiple nuclear bag type fibers in a spindle may receive selective innervation from multiple efferent axons or may all be controlled by the same fusimotor neuron. In some cases one of the nuclear bag fibers in a spindle is operated by an axon that also operates the bundle of nuclear chain fibers, and the other nuclear bag fiber in the same spindle is independently controlled. In most cases, the nuclear chain fibers in a spindle receive innervation from one to several axons independent of the nuclear bag fibers, but in a minority, the nuclear chain fibers share their innervation with one of the nuclear bag fibers as stated above [30].

2.1.2.4 Physiology of Muscle Spindles

Muscle spindles are sensory end organs that detect information on the extrafusal muscles such as contraction speed, changes in muscle length, and changes in contraction speed. Information gathered through the three types of intrafusal muscle fibers and their associated sensory nerve endings are integrated and related to the brain. In general, response of a single-sensory end organ to a defined stimulus is transmitted along the afferent nerve fiber in the form of a series of individual action potentials, each of fixed size, whose rate of occurrence varies according to the strength of the input stimulus [31]. The same is true of afferents from muscle spindles. The responses of sensory primary endings are known to have dynamic (phasic) and static (tonic) components. The former is the response to mechanical stimulus changing with time, such as increasing length of the intrafusal muscle fiber. The latter is the response evoked by the muscle fiber stretched to and maintained at a certain length. Since these intrafusal muscle fibers also receive their own motor innervation, independent adjustments of phasic or tonic aspects of the sensory responses may be possible [30]. That is, the tension or length of the extrafusal muscle fibers surrounding the muscle spindles is not directly reflected on the tension or length of the intrafusal muscle fibers, because the intrafusal muscle fibers have independent motor control allowing the effects of extrafusal muscle contraction on the sensory output to be modified. Based on physiological data, Hunt [26] reported that primary sensory endings show a high dynamic sensitivity in their discharge in response to stretch as well as a maintained static discharge, and the receptor potential has a prominent dynamic component besides a static one. Secondary sensory endings exhibit less dynamic sensitivity in both their discharge in response to stretch and receptor potentials.

2.1.3 Muscle Spindles in the Larynx

Perhaps because fresh human larynges are difficult to come by and because the distribution of muscle spindles is much more sparse in the intrinsic laryngeal muscles compared to other striated muscles, much is still unclear about muscle spindles in the human larynx. Basic functions of the larynx such as respiration, deglutition, and vocalization involve glottic closure, the control of which muscle spindles are an essential factor. However, knowledge of the actual functions of the muscle spindles in the larynx is limited.

Through studies by researchers such as Paulsen [5], Baken [2], Grim [7], Katto [8], and Tamura and Koike [9], it is known that muscle spindles exist in all the intrinsic muscles. According to the criteria proposed by Cooper [10] based on the number of muscle spindles per gram muscle, the intrinsic laryngeal muscles are muscles with a sparse distribution of muscle spindles. There is controversy concerning the difference in distribution of muscle spindles among the intrinsic laryngeal muscles, but most studies agree on the relatively high distribution of muscle spindles in the interarytenoid and posterior cricoarytenoid muscles.

Grim [7] reported that 20 % of the muscle spindles in the posterior cricoarytenoid muscle are single types, while compound types with several parallel spindles together have also been reported by Katto and Okamura [33]. The muscle spindles seen in the intrinsic laryngeal muscles are about 35–150 μm wide and 960–1800 μm long, which are somewhat smaller than muscle spindles seen in other human skeletal muscles. The number of intrafusal muscle fibers in each spindle is one to eight which is fewer than in other human muscles. Human muscle spindles in the interarytenoid muscle had more nuclear chain fibers than nuclear bag type fibers as is known in muscle spindles in general [33]. Nuclear bag type intrafusal muscle fibers in human posterior cricoid muscles were about 13–17 μm in diameter, while nuclear chain fibers had diameters of 7–12 μm [7]. Through histochemical studies, Tamura and Koike [32] showed that the three types of human intrafusal muscle fibers described through histochemical studies in other species, namely, nuclear bag_1 fibers, nuclear bag_2 fibers, and nuclear chain fibers, could be distinguished in human laryngeal intrafusal muscles as well.

Katto [8] studied the structure of muscle spindles in human intrinsic laryngeal muscles and confirmed irregular sensory nerve endings surrounding the intrafusal muscle fibers as flower-spray endings. Peripheral axons penetrating deeply into the sarcoplasm were also seen.

Reports on muscle spindles in the extrinsic laryngeal muscles are relatively sparse. Muscle spindles have been studied in the rat sternothyroid muscle [34], monkey inferior pharyngeal constrictor muscle [35], and human [36] and rabbit [37] digastric muscles (Fig. 2.2).

The function of the muscle spindles in the larynx is not fully understood. Fine control of the vocal cords is necessary to achieve appropriate glottic closure during phonation, and a high level of muscle coordination should be necessary during respiration or deglutition. However, the very low density of muscle spindle distribution in the larynx is inproportionate with the delicate muscle control expected. Therefore, the muscle spindles may not be the only major source of information on muscle tension, but may work with some other receptors of mechanical stimuli in laryngeal control.

2.2 Intramuscular Ganglion

2.2.1 Introduction

The existence of parasympathetic ganglia in the Auerbach plexus between the smooth muscle layers of the intestine is well known, but the distribution of neuronal cell bodies in striated muscle is limited. Some of the few existing early reports are of neurons in the tongue [39–43] and our report of neurons in the inferior pharyngeal constrictor muscle [44].

In 1994, Neuhuber et al. [45] reported that NADPH-diaphorase (NADPHd)-positive neuronal cell bodies were seen in the intrinsic laryngeal muscles during their study of the rat esophagus, but details were not reported presumably because they were not the focus of the study. The innervation of the larynx is composed of a complicated mixture of sensory, motor, and autonomic nerve fibers reaching the larynx via the superior and inferior laryngeal nerves [46–48]. The intrinsic laryngeal muscles are finely controlled through the innervation in connection with respiration, deglutition, and vocalization, and the role the intramuscular ganglion neurons play in this control is a topic of interest. We have studied the intramuscular neurons in the canine larynx.

2.2.2 Distribution of nNOS in the Intramuscular Neurons

Tissue sections prepared from canine intrinsic laryngeal muscles were visualized by NADPHd histochemistry, and the NADPHd-positive neurons were observed and counted

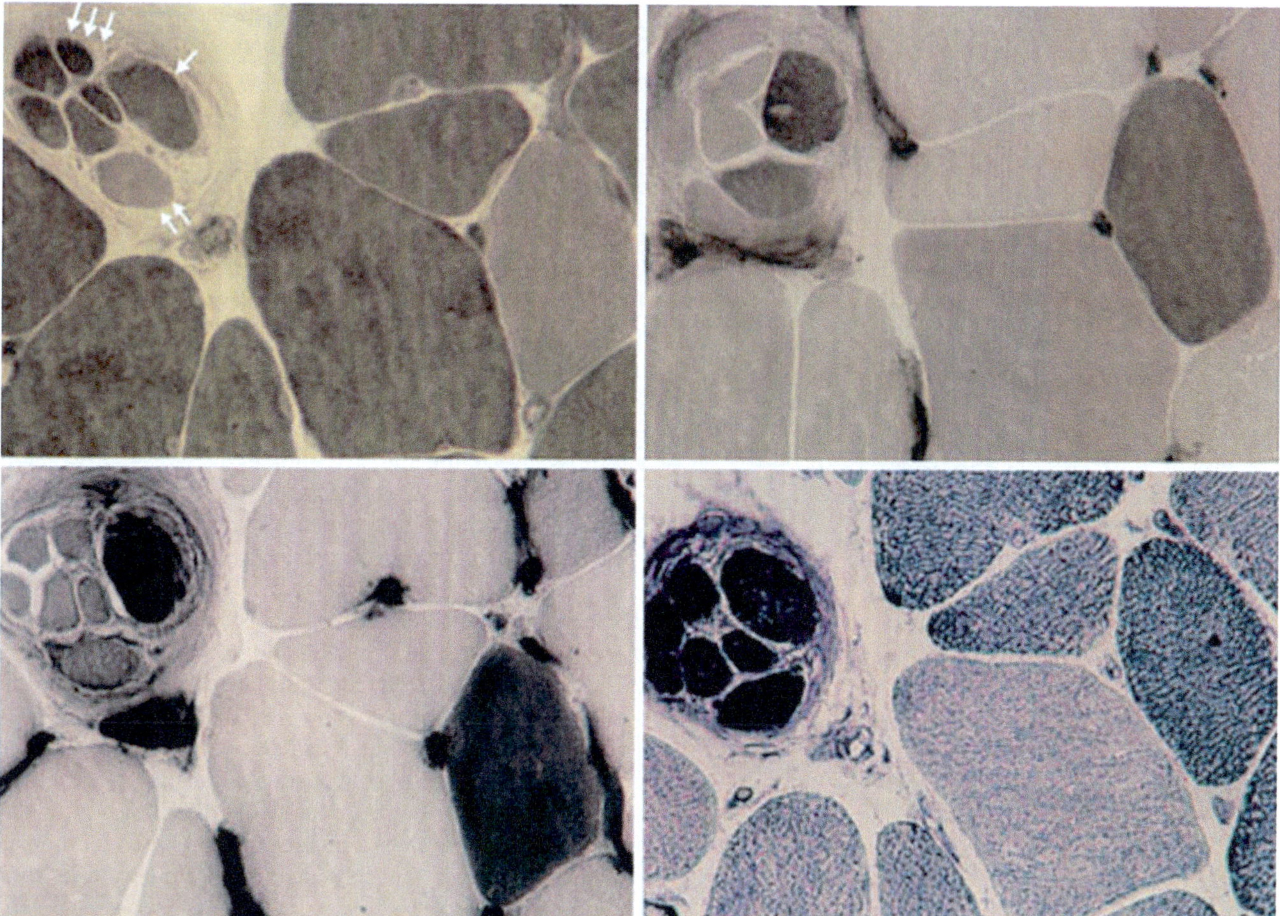

Fig. 2.2 Muscle spindle in cat sternohyoid muscle

(unpublished data). In a subset of the sections, immunofluorescence histochemistry for neuronal nitric oxide synthase (nNOS) was performed before NADPHd histochemistry, and the results were compared.

In all the intrinsic laryngeal muscles studied, neuronal cell bodies with positive stain for NADPHd histochemistry were seen between the muscle layers. Most of the NADPHd-positive neurons were aggregated as a ganglion, but the number of positive neurons in each ganglion varied, and solitary neuronal cell bodies were also seen. Three levels of intensity of stain by NADPHd histochemistry, intense, intermediate, and negative, could be identified in the neurons. The axons of the neurons were clearly visible in the NADPHd intensely stained neurons, and the neurons were bipolar or pseudounipolar in shape (Fig. 2.3). The number of intramuscular ganglia containing NADPHd-positive neurons varied among the intrinsic laryngeal muscles. Many ganglia were seen in the cricothyroid and posterior cricoarytenoid muscles, a few in the lateral cricoarytenoid muscle, and only a very small number of ganglia were seen in the arytenoid and thyroarytenoid muscles (Table 2.1). Most of the intramuscular ganglia were small ganglia consisting of only a few neurons regardless of NADPHd reactivity. Although the number of NADPHd-positive neurons in each intramuscular ganglion was also varied, ganglia with one NADPHd-positive neuron (about 65 % of total ganglia seen), two NADPHd-positive neurons (18 % of ganglia seen), or three NADPHd-positive neurons (7 % of ganglia seen) were the most common and ganglia with no positive neurons at all were also seen. There was no evident localization in the distribution of NADPHd-positive neurons within each intrinsic laryngeal muscle with the exception of the thyroarytenoid muscle, in which case the few neurons seen were in the lateral part of the muscle and no neurons were seen in the medial "vocalis" part of the muscle. Large ganglia containing multipolar NADPHd-positive neurons were seen along thick bundles of nerve fibers running across the muscle layer, but these ganglia were excluded from the count. Such large ganglia were considered to be parasympathetic intralaryngeal ganglia, which will be addressed in another chapter (▶ see Chap. 7). In the sections double stained for NADPHd histochemistry and nNOS immunohistochemistry, the results of the two staining methods matched well (Fig. 2.4).

Since the results of NADPHd histochemistry and nNOS immunohistochemistry were identical, NADPHd histochemistry is a reliable marker for nNOS in the larynx, and positive neurons may synthesize nitric oxide (NO) as a neurotransmitter. The striated muscles of the intrinsic laryngeal muscles are nNOS negative [49], so the neurons that were stained darker than the surrounding muscle tissue were counted as positive cells in the study.

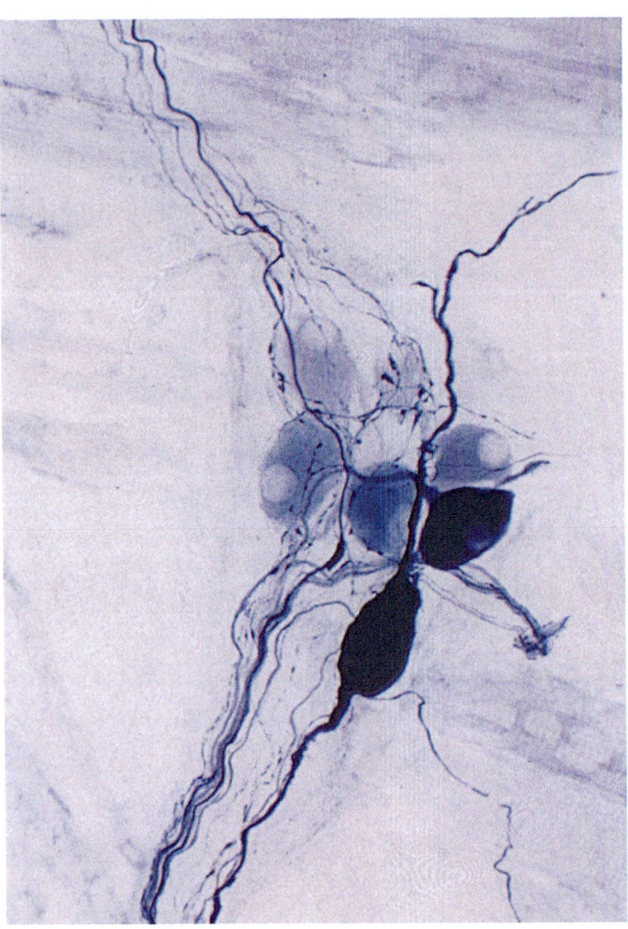

Fig. 2.3 Canine intramuscular ganglion with bipolar and pseudounipolar NADPHd-positive neurons (Revised from [58])

Table 2.1 Number of intramuscular ganglia in each intrinsic laryngeal muscle

Cricothyroid	26.2 ± 11.3
Posterior cricoarytenoid	18.3 ± 2.1
Lateral cricoarytenoid	7.4 ± 4.9
Thyroarytenoid	7.9 ± 3.5
Arytenoid	1.0 ± 0.7

N = 5 dogs, average ± S.D.

Nitric oxide is known to participate in the sympathetic, parasympathetic, and sensory innervation as a neurotransmitter in the central and peripheral nervous system [50–53]. However, it has been generally accepted that nNOS does not exist in cranial or spinal motor neurons unless induced by specific stimulation such as injury [54, 55]. In striated muscle in the rat esophagus, nNOS-positive parasympathetic motor innervation from the myenteric plexus has been reported to exist [45]. We did not find nerve fibers from the intramuscular neurons innervating the intrinsic laryngeal muscles through neuromuscular junctions in our study. Therefore, the NADPHd-positive intramuscular neurons are more probably parasympathetic or sensory than motor in function.

There have been several reports of ganglia existing in the larynx, some as early as the report by Elze [56] in 1903. Most of the detailed descriptions such as on distribution, size, or number in the study were of intralaryngeal ganglia (▶ see Chap. 7 on intralaryngeal ganglia) or similar ganglia and not of neurons in the intrinsic laryngeal muscle layers. The earliest report we were able to find on the "intramuscular ganglion" in the intrinsic laryngeal muscles was by Geronzi [57]. However, the ganglia reported in the posterior cricoarytenoid and cricothyroid muscles were situated along nerve bundles and may have been "intralaryngeal" ganglia found along branches of the superior or inferior laryngeal nerve as it runs between the muscle tissues. In the same report, nerve fibers originating from the ganglionic neurons were seen reaching the muscle end plates [57], but no later report has confirmed the existence of such fibers. Intramuscular neurons did not appear in reports for some time, and a report by Neuhuber et al. [45] in 1994 described neurons in the esophageal muscle layers in detail, but did not focus on the larynx. The first detailed study on intramuscular ganglia in the intrinsic laryngeal muscles and their ganglionic neurons was reported by the authors [58] in 1996.

The neurons of parasympathetic ganglia are generally multipolar. Since the neuronal cell bodies observed in the intrinsic laryngeal muscles were bipolar or pseudounipolar in shape, which is a morphology more common in sensory neurons, a sensory nature of these neurons is suggested. Contrary to this, the large ganglia with multipolar neurons seen along thick nerve bundles between muscle layers in our study were probably parasympathetic ganglia in the periphery such as intralaryngeal ganglia. The fine structure of intramuscular neurons and fibers in the rat posterior cricothyroid muscle studied by electron microscope has been reported [59]. Solitary ganglion cells distributed among muscle fibers and near the bifurcation of arterioles were seen together with nonmyelinated nerve fibers that often formed synapses with each other and with the cell body of ganglion cells. Mainly clear and spherical synaptic vesicles of approximately 50 nm in diameter were seen in the nonmyelinated nerve fibers, and so the nerve fibers were considered to be cholinergic, although the origin of the fibers was not described.

The distribution of muscle spindles in the intrinsic laryngeal muscles is not even [60–62]. Baken and Noback [62] reported that human cricothyroid muscles lack muscle spindles and that the muscle spindles in the thyroarytenoid muscles were all in the medial part of the muscle. Although there may be a species difference, our observations that many intramuscular ganglia are seen in the cricothyroid muscle and none seen in the medial part of the thyroarytenoid muscle show that the distribution may be complementary, and intramuscular ganglia may participate in proprioception in muscle tissue that have less information from muscle spindles. Since neurons in the medial part of the thyroarytenoid muscle within strongly vibrating tissue would be constantly

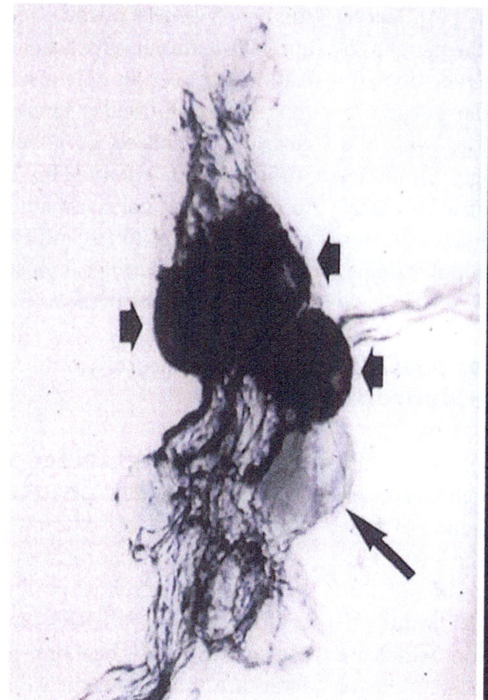

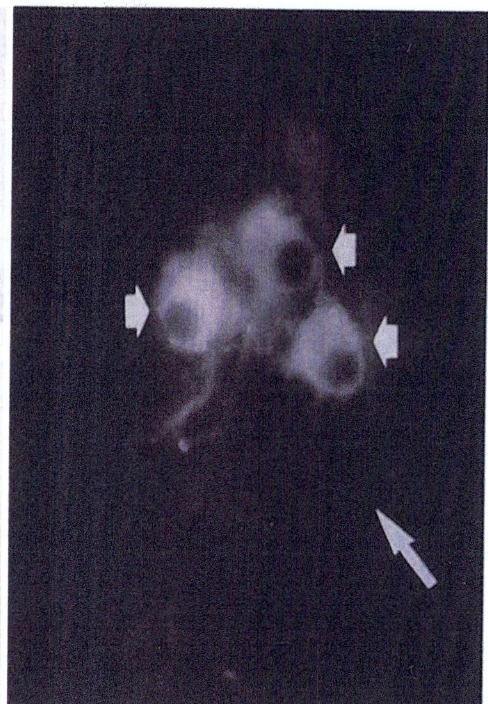

Fig. 2.4 Results of NADPHd histochemistry (left) and nNOS immunofluorescence histochemistry (*right*) in the dog larynx. Results match well (Revised from [58])

stimulated during vocalization, it would not be surprising that intramuscular ganglia were not seen in the medial part of this muscle if the ganglia are sensory.

2.2.3 Neurotransmitters in the Intramuscular Ganglia in the Intrinsic Laryngeal Muscles

Immunohistochemistry using anti-vasoactive intestinal polypeptide (VIP), anti-calcitonin gene-related peptide (CGRP), anti-tyrosine hydroxylase (TH), and anti-heme oxygenase-2 (HO-2) antibodies was utilized to study the canine cricothyroid and posterior cricoarytenoid muscles [63]. These two muscles were selected because previous studies showed that ganglia were most abundant in these muscles as stated above.

No neurons and fibers positive for TH were seen in intramuscular ganglia in the intrinsic laryngeal muscles. Neuronal cell bodies positive for CGRP were not seen, but CGRP-positive nerve fibers were found to exist within the ganglia. Intramuscular ganglia with VIP-positive neurons and intramuscular ganglia with HO-2-positive neurons were seen (Fig. 2.5), but the numbers of such ganglia were much fewer than intramuscular ganglia with NADPHd-positive neurons. There was no difference in these results between the cricothyroid and posterior cricoarytenoid muscles. In the cricothyroid muscle, large ganglia with VIP-positive neurons were seen along thick nerve bundles running across the muscle, similar to the intralaryngeal ganglia seen by NADPHd histochemistry. The multipolar shape of the neurons could be clearly seen as in the case of NADPHd histochemistry.

There are reports on intramuscular ganglia in the tongue of several species, and based on observations using histochemical or electron microscopic methods, these ganglia have been considered parasympathetic [39–43]. A transection experiment in the frog glossopharyngeal nerve has shown that neurons found within the nerve bundle of the glossopharyngeal nerve are parasympathetic postganglionic neurons [43]. The small 8–10 μm neurons Fitzgerald and Alexander reported along nerve bundles in the tongue were also multipolar cells [40]. Therefore the intramuscular neurons in the intrinsic laryngeal muscles differ from these neurons in shape. The neurons in the muscles of the tongue may be equivalent to the multipolar neurons found along thick nerve bundles in the larynx or neurons in the intralaryngeal ganglia.

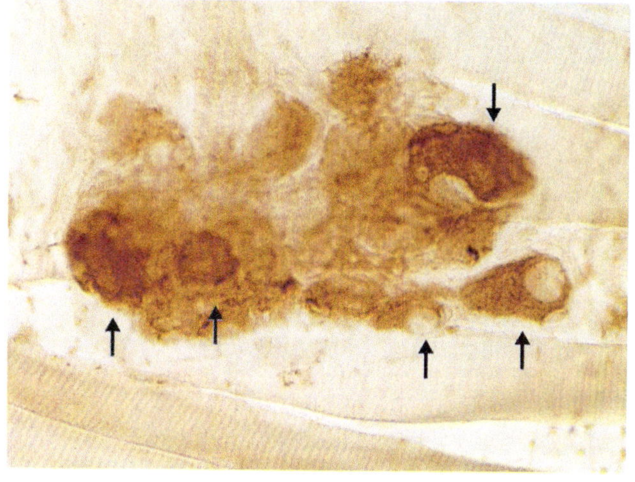

Fig. 2.5 Intramuscular ganglion with HO-2-positive (*arrows*) and HO-2-negative neurons in the canine cricothyroid muscle (Revised from [63])

We have reported the existence of intramuscular ganglia in the canine cricopharyngeal muscle with many VIP-positive neurons, but no CGRP- or TH-positive neurons [44]. Nerve fibers positive for VIP or CGRP were seen in the ganglia in the cricopharyngeal muscle. Therefore, from the types of neurotransmitters seen in the ganglia, intramuscular ganglia in the intrinsic laryngeal muscles have similarities with the intramuscular ganglia in the cricopharyngeal muscle. However, the intramuscular ganglia in the intrinsic laryngeal muscles differ in that they are small ganglia with few neurons and that the neurons have a bipolar or pseudounipolar shape.

2.2.4 Coexistence of NADPHd, VIP, and HO-2 in the Intramuscular Ganglion Neurons

Double-stain techniques by performing immunohistochemistry for VIP or HO-2 followed by NADPHd histochemistry on the same sections were used to study the colocalization of these neurotransmitters (unpublished data). Some of the NADPHd-positive neurons in the intramuscular ganglia in the intrinsic laryngeal muscles were also VIP positive; thus, nNOS and VIP coexist in some of the neurons (Fig. 2.6). Neurons which are VIP positive but NADPHd negative were also seen. None of the neurons were positive for both NADPHd histochemistry and HO-2 immunohistochemistry. Since VIP and NO colocalization was seen in only a subset of the neurons, it is evident that the intramuscular ganglia in the intrinsic laryngeal muscles are composed of a heterogeneous group of neurons. Although NO, VIP, and HO-2 are all neurotransmitters known to exist in the parasympathetic system, the lack of neurons colocalizing NADPHd and HO-2 in the intramuscular ganglia in the intrinsic laryngeal muscles shows that they may differ in function from the neurons in the parasympathetic intralaryngeal ganglia because colocalization of NADPHd and HO-2 is seen in some neurons of the latter ganglia.

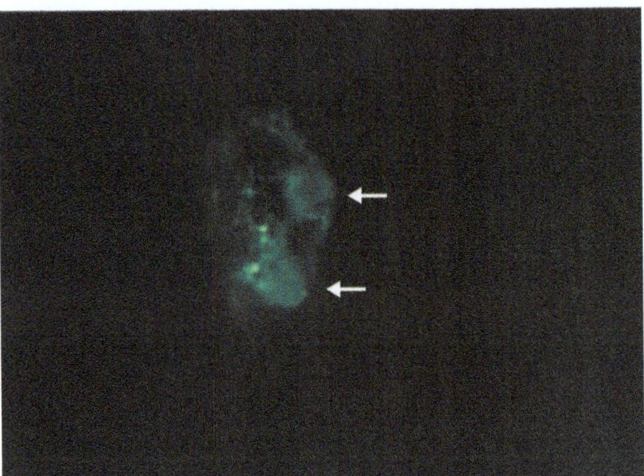

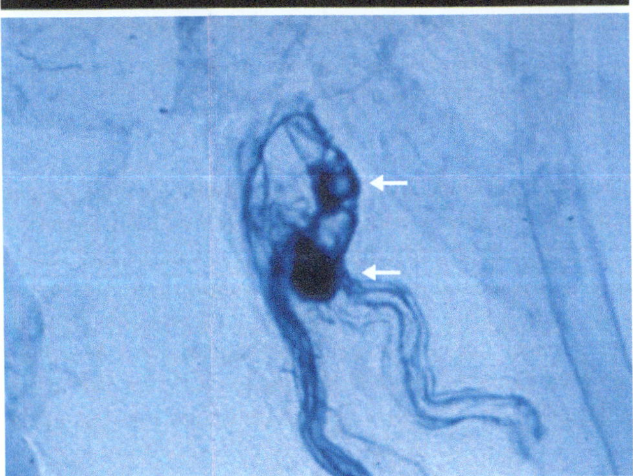

Fig. 2.6 VIP immunohistochemistry (*above*) and NADPHd histochemistry (*below*) in the canine intrinsic laryngeal muscles. Neurons positive for both VIP and NADPHd (*arrows*) are seen in the intramuscular ganglion

2.2.5 Capsaicin Receptors in the Intramuscular Ganglion Neurons

The existence of neurons with the nociceptive receptor transient receptor potential vanilloid 1 (TRPV1) was studied in the rat [64] (Fig. 2.7). TRPV1 is a capsaicin receptor channel, but is also activated by heat or protons [65, 66], and various modulators are known to change the threshold of its response to heat [67–70]. As a polymodal receptor, TRPV1 has the ability to integrate various inputs into the neuronal output. The coexistence of TRPV1 and nNOS was seen in some of the neurons of the intramuscular ganglia [64]. As a gaseous neurotransmitter, NO can diffuse quickly to the surrounding tissue and provide rapid influence such as control of vascular tension and glandular secretion during inflammation. The coexistence of TRPV1 and nNOS in neurons of the intramuscular ganglia in the intrinsic laryngeal muscles would enable such neurons to integrate multiple inputs from neuronal sources and from surrounding tissue and provide a rapid response in the periphery. However, since a location closer to large vessels and glands would be more effective in autonomic control of peripheral tissue, the position of the intramuscular ganglia within the intrinsic laryngeal muscles may be more suited for a sensory function, but with parasympathetic local response capabilities. The existence of TRPV1 in only a subset of the neurons in the intramuscular ganglia again shows their heterogeneity.

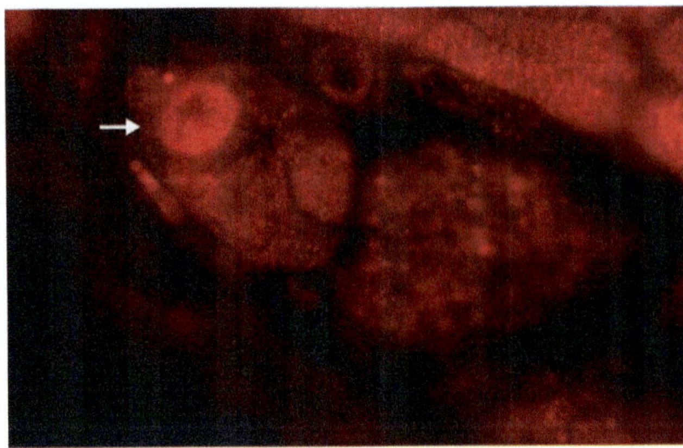

Fig. 2.7 TRPV1-positive neuron (*arrow*) seen in an intramuscular ganglion in the rat (Revised from [64])

References

1. Kerschner L. Beitrag zur Kenntnis der sensiblen Endorgane. Anat Anz. 1888;3:288–96.
2. Ruffini A. Osservazioni critiche allo studio del Dott. Andrea Rossi sulle terminazioni nervose di senso della pelle dell'uomo. Arch Internaz. 1893; 11:16.
3. Sherrington CS. On the anatomical constitution of nerves on skeletal muscles: with remarks on recurrent fibres in the ventral spinal nerve roots. J Physiol. 1894;17:211–58.
4. Goerttler K. Die Anordnung, Histologie und Histogenese der quergestreiften Muskulature in menschlichen Stimmband. Z Anat Entw Gesh. 1950;115:352–401.
5. Paulsen K. Untersuchungen uber das Vorkommen und die Zahl von Muskelspindeln im M. vocalis des Menschen. Z Zellforsch. 1958;47:363–6.
6. Baken RJ. Neuromuscular spindle in intrinsic muscle of human larynx. J Speech Hear Res. 1971;14:513–8.
7. Grim M. Muscle spindles in the posterior cricoarytenoid muscle of the human larynx. Folia Morphol. 1967;15:124–31.
8. Katto Y. Muscle spindle in intrinsic muscles of human larynx: distribution and morphological features. Nippon Jibiinkoka Gakkai Kaiho. 1987;90:59–67.
9. Hirayama M, Matsui T, Tachibana M, Ibata Y, Mizukoshi O. An electron microscopic study of the muscle spindle in the arytenoid muscle of the human larynx. Arch Otorhinolaryngol. 1987;244:249–52.
10. Cooper S. Muscle spindles and other muscle receptors. In: Bourne GH, editor. The structure and function of muscle. New York: Academic; 1960. p. 381–420.
11. Barker D. Morphology of muscle receptors. In: Hunt CC, editor. Handbook of sensory physiology vol 3, pt 2. Berlin: Springer-Verlag; 1974. p. 1–190.
12. Boyd IA. The motor innervation of mammalian muscle spindles. J Physiol. 1961;159:7–9.
13. Boyd IA. The structure and innervation of the nuclear bag muscle fibre system and the nuclear chain fibre system in mammalian muscle spindles. Proc R Soc. 1962;B245:81–136.
14. Spiro AJ, Beilin RL. Human muscle spindle histochemistry. Arch Neurol. 1969;20:271–5.
15. Kucera J, Dorovini-Zis K. Types of human intrafusal muscle fibers. Muscle Nerve. 1979;2:437–51.
16. Sahgal V, Morgen CA. Histochemical and morphological changes in human muscle spindle in upper and lower motor neuron lesions. Acta Neuropathol (Berl). 1976;34:41–6.
17. Yellin H. A histochemical study of muscle spindles and their relationship to extrafusal fiber types in the rat. Am J Anat. 1969;125:31–46.
18. Barker D, Stacey MJ. Rabbit intrafusal muscle fibres. J Physiol. 1970;210:70–2.
19. Ovalle WK, Smith RS. Histochemical identification of three types of intrafusal muscle fibres in the cat and monkey based on the myosin ATPase reaction. Can J Physiol Pharmacol. 1972;50:195–202.
20. Saito M, Tomonaga M, Hirayama K, Narabayashi H. Histochemical study of normal human muscle spindle. Histochemical classification of intrafusal muscle fibers and intrafusal nerve endings. J Neurol. 1977;216:79–89.
21. Ogata T, Mori M. Histochemical study of oxidative enzymes in vertebrate muscles. J Histochem Cytochem. 1964;12:171–82.
22. Banks RW, Harker DW, Stacey MJ. A study of mammalian muscle fibres using a combined histochemical and ultrastructural technique. J Anat. 1977;123:783–96.
23. Guth L, Samaha FJ. Procedure for the histochemical demonstration of actomyosin ATPase. Exp Neurol. 1970;28:365–7.
24. Bewick GS, Banks RW. Mechanotransduction in the muscle spindle. Special issue on physiological aspects of mechano-sensing. Eur J Physiol. 2014. doi:10.1007/s00424-014-1536-9.
25. Matthews PB. The static and dynamic fusimotor fibres – their present status. Bull Schweiz Akad Med Wiss. 1971;27:235–55.
26. Hunt CC. The physiology of muscle receptors. In: Hunt CC, editor. Handbook of sensory physiology vol3, pt2. Berlin: Springer-Verlag; 1974. p. 191–234.
27. Boyd IA, Gladden MH. Morphology of mammalian muscle spindles, a review. In: Boyd IA, Gladden MH, editors. The muscle spindle. London: Macmillan; 1985. p. 3–22.
28. Laporte Y. The motor innervation of the mammalian muscle spindle. In: Porter R, editor. Studies in neurophysiology. Cambridge: Cambridge University Press; 1978. p. 45–59.
29. Barker D, Emonet-Denand F, Laporte Y, Stacey MJ. Identification of the intrafusal endings of skeletofusimotor axons in the cat. Brain Res. 1980;185:227–37.
30. Boyd IA, Ward J. Motor control of nuclear bag and nuclear chain intrafusal fibres in isolated living muscle spindles from the cat. J Physiol. 1975;244:83–112.
31. Adrian ED, Zotterman Y. The impulses produced by sensory nerve endings: part II. The response of a single end-organ. J Physiol. 1926;61:151–71.
32. Tamura K, Koike Y. Enzyme-histochemistry of muscle spindles in human intrinsic laryngeal muscles. Jibiinkoka Rinsyo (Supp). 1999;101:45–9.
33. Kattou Y, Okamura H. Morphological study of muscle spindles in arytenoid muscle of human larynx. Nippon Jibiinkoka Gakkai Kaiho. 1983;86:661–5.
34. Andrew BL. The respiratory displacement of the larynx: a study of innervation of accessory respiratory muscles. J Physiol. 1955;130:474–87.
35. Sengupta BN, Sengupta S. Muscle spindles in the inferior constrictor pharyngis muscle of the crab-eating monkey (Macaca irus). Acta Anat (Basel). 1978;100:132–5.
36. Lennartsson B. Muscle spindles in the human anterior digastric muscle. Acta Odontol Scand. 1979;37:329–33.
37. Muhl ZF, Kotov O. Muscle spindles in the digastric muscle of the rabbit. J Dent Res. 1988;67:1243–5.
38. Sano Y. Shinkei-Kagaku (Keitaigakuteki Kiso) Vol.1 Neuron and Glia. Kyoto: Kinpodo; 1995. pp. 475–86.
39. Gairns FW, Garven HSD. Ganglion cells in the mammalian tongue. J Physiol Lond. 1952;118:53–4.
40. Fitzgerald MJ, Alexander RW. The intramuscular ganglia of the cat's tongue. J Anat. 1969;105:27–46.
41. Bhargava KN, Bhargava AK. Some anatomical and neurohistochemical observations on the chick tongue. Mikroskopie. 1974;30:193–201.
42. Baecker B, Yanaihara N, Forssmann WG. VIP innervation of the tongue in vertebrates. Anat Embryol (Berl). 1983;167:173–89.
43. Inoue K, Kitada Y. Parasympathetic postganglionic cells in the glossopharyngeal nerve trunk and their relationship to unmyelinated nerve fibers in the fungiform papillae of the frog. Anat Rec. 1991;230:131–5.
44. Tadaki N, Hisa Y, Uno T, Koike S, Okamura H, Ibata Y. Neurotransmitters for the canine inferior pharyngeal constrictor muscle. Otolaryngol Head Neck Surg. 1995;113:755–9.
45. Neuhuber WL, Worl J, Berthoud HR, Conte B. NADPH-diaphorase-positive nerve fibers associated with motor endplates in the rat esophagus: new evidence for co-innervation of striated muscle by enteric neurons. Cell Tissue Res. 1994;276:23–30.
46. Wyke BD, Kirshner JA. Neurology of the larynx. In: Hinchcliffe R, Harrison D, editors. Scientific foundation of otolaryngology. London: Heinemann; 1976. p. 546–74.
47. Hisa Y. Fluorescence histochemical studies on the noradrenergic innervation of the canine larynx. Acta Anat. 1982;113:15–25.
48. Wallach JH, Rybicki KJ, Kaufman MP. Anatomical localization of the cells of origin of efferent fibers in the superior laryngeal and recurrent laryngeal nerves of dogs. Brain Res. 1983;261:307–11.
49. Hisa Y, Tadaki N, Uno T, Koike S, Tanaka M, Okamura H, Ibata Y. Nitrergic innervation of the rat larynx measured by nitric oxide synthase immunohistochemistry and NADPH-diaphorase histochemistry. Ann Otol Rhinol Laryngol. 1996;105:550–4.
50. Modin A, Weitzberg E, Hokfelt T, Lundberg JM. Nitric oxide synthase in the pig autonomic nervous system in relation to the influence of NG-nitro-L-arginine on sympathetic and parasympathetic vascular control in vivo. Neuroscience. 1994;62:189–203.

51. Holzer P, Wachter C, Heinemann A, Jocic M, Lippe IT, Herbert MK. Sensory nerves, nitric oxide and NANC vasodilatation. Arch Int Pharmacodyn. 1995;329:67–79.
52. Papka R, McNeill D, Thompson D, Schmidt H. Nitric oxide nerves in the uterus are parasympathetic, sensory, and contain neuropeptides. Cell Tissue Res. 1995;279:339–49.
53. Xu Z, Li P, Tong C, Figueroa J, Tobin JR, Eisenach JC. Location and characteristics of nitric oxide synthase in sheep spinal cord and its interaction with alpha(2)-adrenergic and cholinergic antinociception. Anesthesiology. 1996;84:890–9.
54. Wu W, Liuzzi FJ, Schinco FP, Depto AS, Li Y, Mong JA, Dawson TM, Snyder SH. Neuronal nitric oxide synthase is induced in spinal neurons by traumatic injury. Neuroscience. 1994;61:719–26.
55. Yu WH. Nitric oxide synthase in motor neurons after axotomy. J Histochem Cytochem. 1994;42:451–7.
56. Elze C. Kurze Mitteilung uber ein Ganglion in Nervus laryngeus superiorus des Menschen. Zeitschr Anat Entwicklungsgesch. 1923;69: 630.
57. Geronzi G. On the presence of nerve ganglia intramuscularly in certain intrinsic muscles of the larynx. Arch Int Laryng. 1904;6: 854–65.
58. Hisa Y, Koike S, Uno T, Tadaki N, Tanaka M, Okamura H, Ibata Y. Nitrergic neurons in the canine intrinsic laryngeal muscle. Neurosci Lett. 1996;203:45–8.
59. Desaki J, Yamagata T, Kawakita S. The distribution of ganglion cells in the posterior cricoarytenoid muscle of the normal adult rat. A light and electron microscopic study. Arch Histol Cytol. 2003;66(1):27–36.
60. Keene MF. Muscle spindles in human laryngeal muscles. J Anat. 1961;95:25–9.
61. Rossi G, Cortesina G. Morphological study of the laryngeal muscles in man. Insertions and courses of the muscle fibers, motor end-plates and proprioceptors. Acta Otolaryngol. 1965;59:575–92.
62. Baken RJ, Noback R. Neuromuscular spindles in intrinsic muscles of a human larynx. J Speech Hear Res. 1971;14:513–8.
63. Koike S, Hisa Y. Neurochemical substances in neurons of the canine intrinsic laryngeal muscles. Acta Otolaryngol. 1999;119:267–70.
64. Koike S, Uno T, Bamba H, Shibata T, Okano H, Hisa Y. Distribution of vanilloid receptors in the rat laryngeal innervation. Acta Otolaryngol. 2004;124:515–9.
65. Caterina MJ, Schumacher MA, Tominaga M, Rosen TA, Levine JD, Julius D. The capsaicin receptor: a heat – activated ion channel in the pain pathway. Nature. 1997;389:816–24.
66. Caterina MJ, Rosen TA, Tominaga M, Brake AJ, Julius D. A capsaicin-receptor homologue with a high threshold for noxious heat. Nature. 1999;398:436–41.
67. Julius D, Basbaum AI. Molecular mechanisms of nociception. Nature. 2001;413:203–10.
68. Mizumura K, Kumazawa T. Modification of nociceptic responses by inflammatory mediators and second messengers implicated in their action- a study in canine testicular polymodal receptors. Prog Brain Res. 1996;113:115–41.
69. Wood JN, Perl ER. Pain. Curr Opin Genet Dev. 1999;9:328–32.
70. Woolf CJ, Salter MW. Neuronal plasticity: increasing the gain in pain. Science. 2000;288:1765–9.

Motor Nerve Endings

Ryuichi Hirota, Shinobu Koike, and Yasuo Hisa

R. Hirota
Hirota ENT Clinic, Nishinoyama Nakatorii-cho, Yamashina-ku, Kyoto 607-8306, Japan
e-mail: rhirota@koto.kpu-m.ac.jp

S. Koike • Y. Hisa (✉)
Department of Otolaryngology-Head and Neck Surgery, Kyoto Prefectural University of Medicine, Kawaramachi-Hirokoji, Kamigyo-ku, Kyoto 602-8566, Japan
e-mail: yhisa@koto.kpu-m.ac.jp

Y. Hisa (ed.), *Neuroanatomy and Neurophysiology of the Larynx*, DOI 10.1007/978-4-431-55750-0_3

3.1 Introduction

A neuron is divided into a region for stimulus acceptance, a region for stimulus transmission, and an effector region. The effector region of the efferent neurons in the somatic nervous system, which are neurons forming part of the peripheral nervous system, is the motor neuron ending, and the nerve fibers end at the motor endplates.

3.2 Motor Nerve Endings

3.2.1 Nerve Fibers and Motor Endplates

Nerve fibers leaving motor neurons reach the surface of muscle fibers (myocytes), lose their myelin sheath, and branch off. Furthermore, these naked axons end at the motor endplates or neuromuscular junction. Motor endplates are chemical synapses made up of plate-shaped α-motor neuron (α-MN) endings, a gap, and the corresponding muscle fibers. Acetylcholine (ACh) is a chemical transmitter in vertebrate animals.

The nerve endings of efferent neuron fibers end on their respective effector organs and are therefore also known as effector organ endings. Motor neuron endings can be classified into those that end at common skeletal muscle (extrafusal skeletal muscle fibers) neuromuscular junctions and those that end on muscle spindle neuromuscular junctions of innervate skeletal muscle fibers (intrafusal muscle fibers) within the muscle spindle. Somatic motor neuron fibers are either large-diameter, myelinated α-motor fibers (12–20 μm) or narrow, myelinated γ-motor fibers (2–8 μm). The former ends on skeletal muscles, while the latter ends on muscle spindles [1].

3.2.2 Methods of Observing Motor Endings

Skeletal muscle motor neuron endings were first observed in 1840 by Doyère (König and von Leden [2]). The existence of motor endplates, which he called *Endplatte* was discovered by Kühne [3]. The structure of motor endplates could be observed with light microscopes by using various methods such as methylene blue staining or the silver impregnation technique. Because abundant quantities of acetylcholinesterase (AChE), which breaks down ACh, are found in the synaptic cleft, enzyme histochemical techniques are used to detect AChE, making it possible to portray the position of the endplate under a light microscope clearly. In addition, the position of the acetylcholine receptor (AChR) on the muscle surface can be identified by using AChR antibodies or snake venom toxins [4]. There have also been recent reports of examination using an antibody to synaptophysin, a synaptic vesicle membrane-related protein [5]. Furthermore, by performing detailed examination using electron microscopes and using scanning electron microscopes [6, 7], the fine structure of subneural apparatus (presynaptic membrane, postsynaptic membrane) has been shown.

3.2.3 Motor Endplate Size and Neuronal Innervation

The size of motor endplates is not uniform and differs based on animal species and skeletal muscle type. In humans, the larger the skeletal muscle is, the larger the motor endplates are. Human motor endplates are usually between 40 and 60 μm in length [8]. There are several reports indicating that the size of motor endplates is related to the diameter of the muscle fibers [9, 10].

One motor neuron will innervate one or more muscle fibers, and the smaller the ratio is, the more accurate and delicate the muscle movements are. In the case of the extraocular muscles or the muscles in the fingers, one axon ends on two to three muscle fibers, while one axon innervates 100–300 muscle fibers in muscles that are required to be strong, such as the gluteus muscles or femoral muscles. One axon supplying the intralaryngeal muscles is said to innervate approximately 30 muscle fibers [11]. In addition, muscle fiber groups that are innervated by individual motor neurons are known as motor units, but rather than individual motor units being independent and innervating individual muscles, one muscle sometimes made up of a mixture of multiple motor units distributed throughout the muscle. If there is one motor endplate present for each muscle fiber, this is known as a single-motor endplate, or focal innervation. If there are multiple motor endplates present on a single muscle fiber, this is known as a multi-motor endplate, or multiple innervations. Reports indicate that multi-motor endplates are present on some muscles such as the extraocular muscles, the orbicularis oculi, and the muscles of the auditory ossicles [12–14].

3.2.4 Structure of the Motor Neuron Endings

These can be classified into three main types based on morphological characteristics observed under a light microscope (▣ Fig. 3.1). The disk-shaped endings are known as *terminaison en plaque* (solid endplates), the endings that resemble bunches of grapes are known as *terminaison en grappe*, and the linear endings are known as *terminaison en ligne. Terminaison en plaque* in particular is said to be the typical motor endplates that form on extrafusal muscle fibers and cause rapid contraction of the entire muscle fiber in response to a transmitted impulse. In the case of *terminaison en grappe*, an impulse will bring about localized excitation of muscles, rather than causing a contraction of the entire muscle. *Terminaison en grappe* endings are commonly observed in the extraocular muscles [1, 15].

When the motor neuron fibers are close to the endings, the myelin sheath disappears, and terminal branches appear and disperse. The terminal branches are covered by Schwann cells, and the Schwann cells that are distributed in the terminal region are known as terminal glia. The branched axon terminal fibers have button-shaped bulges at the end, but the synapse is formed by a common surface made up of the axon membrane of the axon terminal and the sarcolemma covering the sarcoplasm. This is known as the primary synaptic

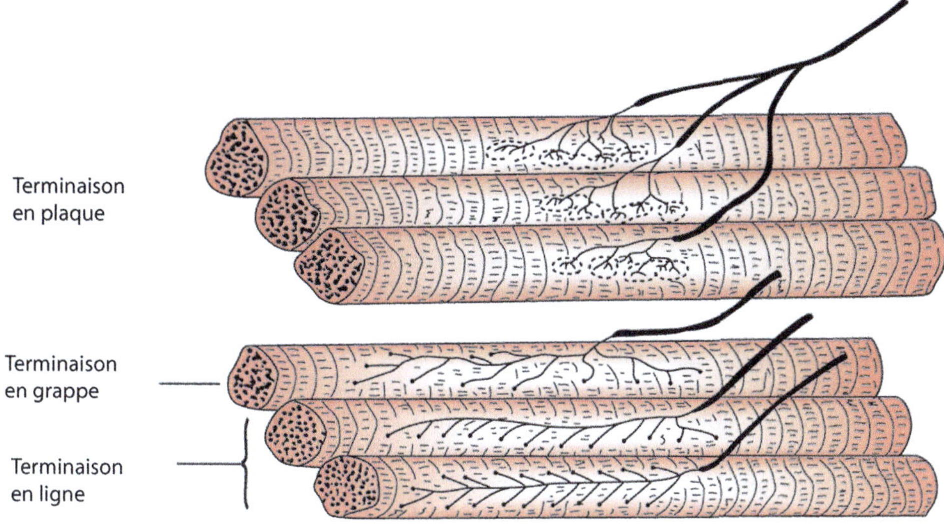

Fig. 3.1 Morphological characteristics of motor endplates

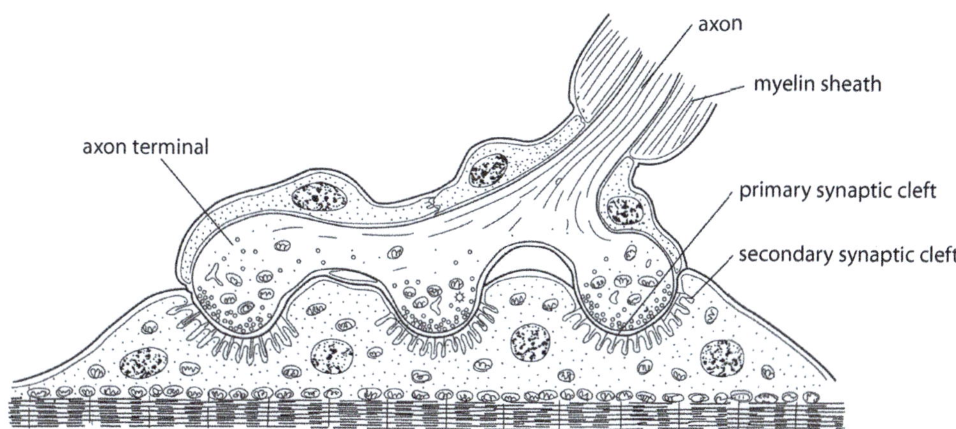

Fig. 3.2 Fine structure of the motor endplates

cleft. The gap formed is approximately 40–60 nm wide and contains a layer of proteins and mucopolysaccharides. The axonal membrane facing this gap (presynaptic membrane) is smooth, but there are several notch-shaped folds in the depressions present in the sarcolemma on the other side (postsynaptic membrane), and these folds are known as the secondary synaptic clefts or junctional folds [1] (Fig. 3.2).

There is an abundance of mitochondria at the axon terminal, and several other opaque synaptic vesicles, with a diameter of between 40 and 50 nm, are also present. These synaptic vesicles store ACh. In the axonal membrane facing the primary synaptic cleft, there is an area covered with patches where a dark substance has accumulated. Couteaux [16] called these areas active zones and verified that the active zones appear at the sites corresponding to the secondary synaptic clefts. Birks et al. [17] then verified that synaptic vesicles were localized in the axon terminals and matched those in the active zone. Heuser et al. [18] used the freeze-fracture method and verified that synaptic vesicles adhere to the presynaptic membrane at these sites and that exocytosis occurs. When neuronal excitation reaches the motor endplate, ACh is released into the synaptic cleft through exocytosis of the synaptic vesicles present in the axon terminals. ACh passes through the basement membrane and reaches the receptors on the postsynaptic membrane. This enhances the permeability of the postsynaptic membrane to sodium (Na^+) ions, resulting in localized depolarization known as an endplate potential. AChR is believed to be present in a region lined by a highly electron-dense material on the internal surface of the junctional folds formed in the postsynaptic membrane.

3.3 Laryngeal Motor Nerve Endings

The activity of the intralaryngeal muscles is controlled by complex sensory and motor neuron systems. Each muscle has a specific function, such as the opening or closing of the glottis or the regulation of vocal cord tension, and, moreover, these muscles must all act in harmony, so we know not only the function, but the individual differences in the distribution pattern of the motor endplates and neuronal innervation.

3.3.1 Laryngeal Muscle Motor Endplates and Neuronal Innervation

Because the intralaryngeal muscles perform finely regulated motor actions, the motor endplates are smaller than the motor endplates on other skeletal muscles, and approximately 20,000–30,000 motor endplates are believed to exit on the thyroarytenoid muscle [8].

Sato et al. [19] studied the intralaryngeal muscles in Japanese monkeys and comparatively investigated the size of the endplates in several muscle fibers. They reported that, compared to the other intralaryngeal muscles, the cricothyroid muscle and posterior cricoarytenoid muscle had thicker muscle fibers and larger endplates. Muscles, such as the extraocular muscles that are required to perform fine movements, are observed to have multiple motor endplates for each muscle fiber. In human intralaryngeal muscle, according to Rossi and Cortesina [20], multi-motor endplates are observed on 70–80 % of the vocal muscle fibers, 50 % of the cricothyroid muscle fibers and external cricoarytenoid muscle fibers, and 5 % of the posterior cricoarytenoid muscle fibers. Meanwhile, Sonesson [21], König and von Leden [2] have published reports that contradict these results. Yoshihara et al. [22] reported that all the posterior cricoarytenoid muscles are innervated by a single neuron, and, currently, there are no corroborating opinions.

In addition, controlled muscle movements require feedback pathways from unique sensory receptors, and, generally, the muscle spindles are seen to take on this central role. However, the distribution frequency of muscle spindles in the intralaryngeal muscles is extremely low [23]. The nerve endings in the posterior cricoarytenoid muscles were observed to have morphological characteristics that differ from those observed in the motor neuron endings present in common skeletal muscles and are believed to be spiral nerve endings, the muscle sensory receptors reported to be present in the extraocular muscles. It has been suggested that control of intralaryngeal muscle activity may be accomplished by the muscle spindles and these spiral nerve endings, as is the case with the extraocular muscles [24].

3.3.2 Morphological Characteristics of the Intralaryngeal Muscle Motor Endplates

Through examination after using silver impregnation techniques, aside from the classical three types present in human posterior cricoarytenoid muscles, namely, the *terminaison en plaque*, *terminaison en grappe*, and *terminaison en ligne*, incomplete types and type switching have been observed. On the other hand, *terminaison en plaque* forms the majority of the motor endings in canine posterior cricoarytenoid muscles and indicates that there may be a diversity of motor endings, depending on the species [25]. Yoshihara et al. [22] reported that most of the endings in human posterior cricoarytenoid muscles are of the *terminaison en plaque* type, but there is no consensus on this point.

During observation of the subneural apparatus in the intralaryngeal muscles of rats using scanning electron microscopes [6, 7], three main types could be classified based on the morphological characteristics. There were small endplates with approximately ten recesses corresponding to the primary synaptic cleft, large endplates with 15 or more of the same recesses, and endplates with recesses resembling intricate structures. Because the small endplates are found on relatively thin muscle fibers (10–20 μm) and the percentage included in each of these muscles roughly corresponded to the proportion of type I muscle fibers, it has been suggested that the differences in the subneural apparatus of the endplates are related to the corresponding muscle fiber type.

3.3.3 Distribution of Motor Endplates on Each Intralaryngeal Muscle

Generally, motor endplates on the skeletal muscle accumulate approximately at the center of the muscle belly, but the motor endplates in the intralaryngeal muscles are not necessarily located at the center of each muscle and various characteristic distributions are observed [26–30]. The distribution pattern cannot be uniform presumably because the intralaryngeal muscles require fine control of muscle activity.

In the posterior cricoarytenoid muscle, which is the only abductor of the glottis, the distribution of motor endplates is arc shaped, with relatively few motor endplates at the top and bottom of the muscle [26]. These are reported to be scattered, instead of in a clear belt-shaped pattern [22]. This muscle runs in an oblique and superolateral direction from the fossa on the surface posterior to the lamina of cricoid cartilage and attaches to the muscular process of the arytenoid cartilage.

Rosen et al. [27] reported that the motor endplates are distributed throughout the whole thyroarytenoid muscle, but there are also reports [5] indicating that 74 % of the motor endplates are present in the middle one-third. They form a belt shape along the center of the muscle fibers in the lateral cricoarytenoid muscle, and the distribution is not different from ordinary skeletal muscles [28]. The endplate distribution pattern has an inverted Y shape in the arytenoid muscle, with a wide belt shape at the top and center of the muscle, but divides obliquely into two as the pattern extends in a downward direction [29].

In the cricoarytenoid muscle, the motor endplates are distributed irregularly in the middle two-third in an anterior-posterior direction [30].

3.3.4 Identification of Neurotransmitters in the Motor Neuron Endings in the Intralaryngeal Muscles

Nitric oxide synthase (NOS) localization was used to investigate the laryngeal nervous system in rats. Results indicated that NOS is present in the laryngeal autonomic, sensory, and motor neurons, elucidating the fact that nitric oxide (NO) participates in the laryngeal nervous system as a neurotransmitter [31]. NADPH-diaphorase (NADPHd) histochemistry, which is a histochemical marker of NOS, was used in the rat thyroarytenoid muscle, and motor neuron endings were clearly observed (◘ Fig. 3.3a). Similarly, calcitonin

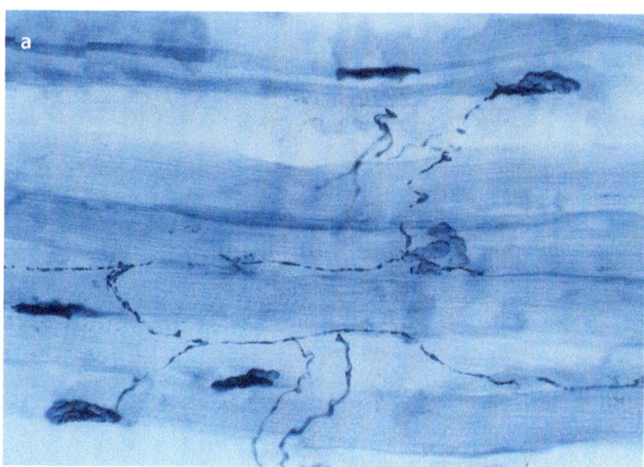

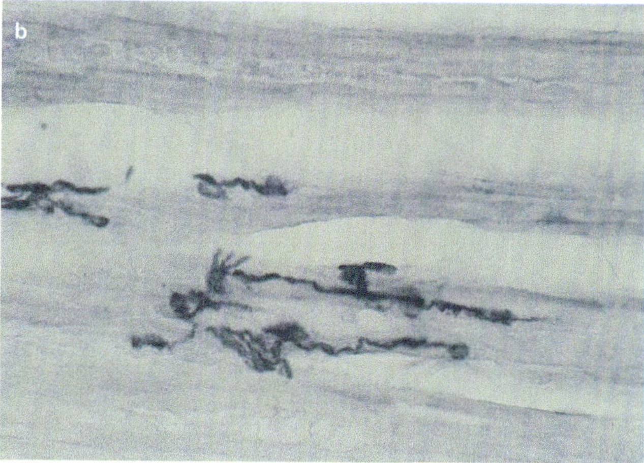

Fig. 3.3 Motor endplates in the thyroarytenoid muscle. (**a**)The NOS tissue marker was used to perform NADPHd histochemistry on rat thyroarytenoid muscle. NADPHd-positive endplate structures are visible. (**b**) CGRP immunohistochemistry was performed on canine thyroarytenoid muscles. The CGRP-positive motor endplates are visible

gene-related peptide (CGRP) was used during localization in canine laryngeal nervous systems. Results of investigations using immunohistochemistry showed that CGRP-positive motor neuron endings were observed [32] (Fig. 3.3b).

There are several unclear points regarding how NO and CGRP are jointly involved as neurotransmitters at motor neuron endings, but Neuuber et al. [33] investigated motor neuron endings in rat esophagus and reported that CGRP-positive neuron endings had thicker axons and larger endplates than NADPHd-positive neuron endings and that NADPhd was present in 3/4 of the subdiaphragmatic esophagus, with CGRP and NADPHd coexisting in only 1/3. There is also control due to a complex innervation during peristalsis in the esophagus, and the involvement of NO and CGRP has been suggested. Similarly, it became possible to elucidate the involvement of NO and CGRP in the complex neuronal regulation present in the intralaryngeal muscles.

3.3.5 Identification of Motor Neuron Endings in the Canine Inferior Constrictor [34]

The thyropharyngeal (thyropharyngeus) and cricopharyngeal (cricopharyngeus) parts of the inferior constrictor lie next to one another in the posterior larynx and play an important role in the second phase of swallowing. Together, the muscles are commonly known as the inferior (pharyngeal) constrictor, but show completely different functions during the process of swallowing. Similar to the middle constrictor, the thyropharyngeus is relaxed at rest, and strong contractions are observed during swallowing. The cricopharyngeus acts as sphincter in the upper gastrointestinal tract and normally remains in a continuous, contracted state in order to prevent the reflux of food, only relaxing during swallowing.

We investigated the neurotransmitters and neuroactive substances in order to understand the anatomical difference between these two nerves. We used gold-labeled cholera toxin (CTBG) as a neuron tracer, and because CGRP, the neurotransmitter at the intralaryngeal muscle motor endplates, was present [32], we investigated the localization of CGRP and AChE, which is observed at the motor endplates.

In the thyropharyngeus, the motor neuron endings that were positive for AChE activity roughly corresponded to the motor neuron endings that were CGRP positive (Fig. 3.4). In the cricopharyngeus, there were multiple motor neuron endings that were positive for AChE activity, but there were quite few that were CGRP positive (Fig. 3.5). In addition, the proportion of CGRP-positive cells that innervate the nucleus ambiguus was 70.3 % in the thyropharyngeus, compared to 21.2 % in the cricopharyngeus (Table 3.1).

The action of CGRP in skeletal muscle cells is well known to be the potentiation of muscle contraction caused by ACh [35] and the regulation of AChR synthesis [36]. Yoshihara et al. [37] showed that ACh and CGRP are both present at the motor neuron endings of the feline posterior cricoarytenoid muscle and reported that CGRP may participate in the regulation of muscle contraction. We presume that CGRP is present in several motor neuron endings in thyropharyngeus, which is required to have powerful contractility during swallowing, and only a few motor neuron endings in the cricopharyngeus, where it is mainly required to regulate the synthesis of AChR.

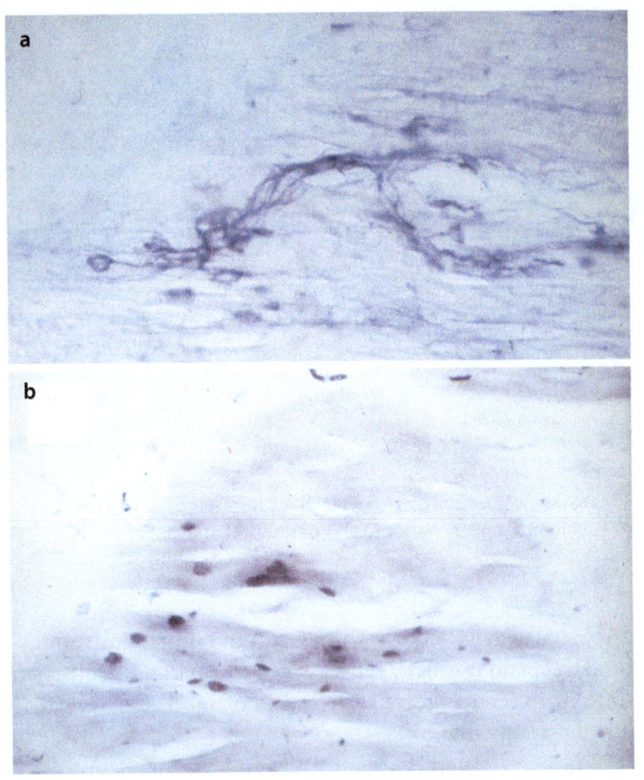

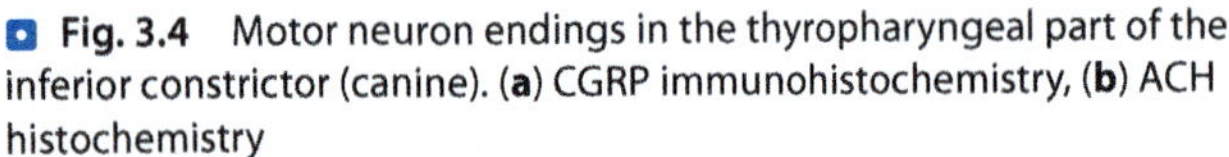

Fig. 3.4 Motor neuron endings in the thyropharyngeal part of the inferior constrictor (canine). (**a**) CGRP immunohistochemistry, (**b**) ACH histochemistry

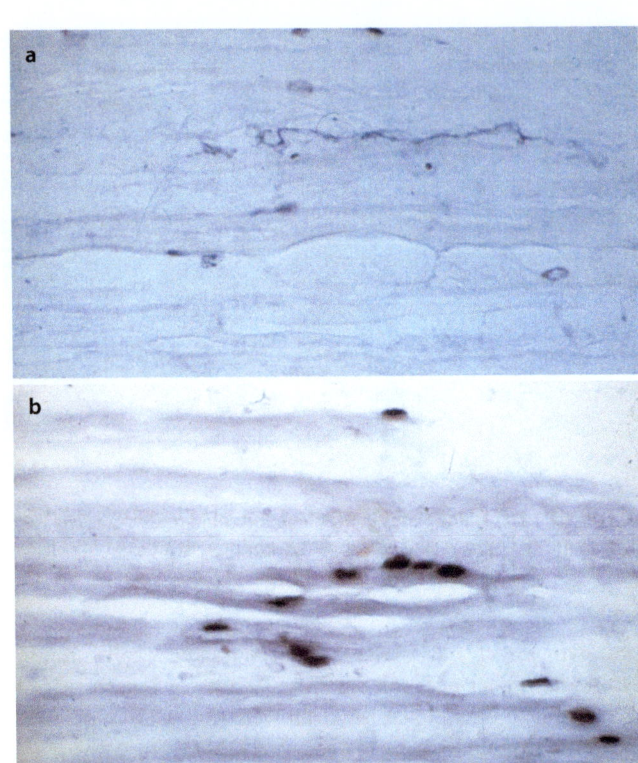

Fig. 3.5 Motor neuron endings in the cricopharyngeal part of the inferior constrictor (canine). (**a**) CGRP immunohistochemistry, (**b**) ACH histochemistry

Table 3.1 CGRP-positive motor neurons in the thyropharyngeal part and cricopharyngeal part of the inferior constrictor (canine) [34]

Animal no.	CTBG positive (A)	CTBG/CGRP positive (B)	B/A (%)
Rate of CGRP-positive neurons among the motor neurons controlled by the nucleus ambiguus in the thyropharyngeal part of the inferior constrictor			
1	118	87	73.7
2	142	88	62.0
3	81	59	72.8
4	78	65	83.3
5	88	56	63.6
Mean	101	71	70.3
Rate of CGRP-positive neurons among the motor neurons controlled by the nucleus ambiguus in the cricopharyngeal part of the inferior constrictor			
1	139	16	11.5
2	124	24	19.4
3	141	19	13.5
4	147	47	32.0
5	136	34	25.0
Mean	137	29	21.2

References

1. Sano Y. Type IV nerve endings and synapses 4. Chemical synapse D. Synapse structure 4) Chemical synapses between neuron effector structures a) Somatic effector organs (motor end plate). Fundamental neuroscience morphology I. Neurons and glia. Kyoto: Kinpodo Publishers; 1995. p. 393–404.
2. König WF, von Leden H. The peripheral nervous system of the human larynx. II. The thyroarytenoid (vocalis) muscle. Arch Otolaryngol. 1961;74:153–63.
3. Kühne W. Über die Endigung der Nerven in den Muskeln. Virchows Arch Pathol Anat. 1863;27:508–33.
4. Ishikawa Y, Shimda Y. Fluorescent staining of acetylcholine receptors at the neuromuscular junction by means of rhodamine labeled erabutoxin b. In: Ito M, Tsukahara N, Kubota K, Yagi K, editors. Integrative control functions of the brain vol III. Kodansha Scientific, Tokyo/Elsevier, Amsterdam; 1981. p 29–31.
5. Sheppert AD, Spirou GA, Berrebi AS, Garnett JD. Three-dimensional reconstruction of immunolabeled neuromuscular junctions in the human thyroarytenoid muscle. Laryngoscope. 2003;113:1973–6.
6. Kawakita S, Yumoto E, Aibara R, Yamagata T, Desaki J. Neuromuscular junctions of the posterior cricoarytenoid muscle in the adult rat: a scanning electron microscopical study. Arch Histol Cytol. 1996;59:375–9.
7. Yamagata T, Kawakita S, Hyodo M, Desaki J. Scanning electron microscopic study of the neuromuscular junctions of the cricothyroid and thyroarytenoid muscles in rats. Acta Otolaryngol. 2000;120:766–70.
8. Coers C. Les variations structurelles normales et pathologiques de la jonction neuro masculaire. Acta Neurol Psychiatr Belg. 1955;55:741–966.
9. Ogata T. A histochemical study on the structural differences of motor endplate in the red, white and intermediate muscle fibers of mouse limb muscle. Acta Med Okayama. 1965;19:149–53.
10. Harris JB, Ribchester RR. The relationship between end-plate size and transmitter release in normal and dystrophic muscles of the mouse. J Physiol. 1979;296:245–65.
11. Williams PL. Gray's anatomy. 38th ed. New York: Churchill Livingstone; 1995. p. 174–81. 806–7, 1251–5, 1637–47, 1723–33.
12. Desmedt JE. Methodes d'etude de la fonction neuromusculaire chez l'homme. Myogramme isometrique, electromyogramme d'excitation et topographie de l'innervation terminale. Acta Neurol Psychiatr Belg. 1958;58:977–1017.
13. Kupfer C. Motor innervation of extraocular muscle. J Physiol. 1960;153:522–6.
14. Fernand VS, Hess A. The occurrence, structure and innervation of slow and twitch muscle fibres in the tensor tympani and stapedius of the cat. J Physiol. 1969;200:547–54.
15. Shimada Y, Ishikawa Y, Kanda T, Yoshihara T. Shinkei-kin Setsugo (Neuromuscular junctions). Saibo (The Cell). 1982;14:304–9.
16. Couteaux R. Moter end plate structure. In: Bourne GH, editor. Structure and function of muscle, vol. 1. New York: Academic; 1960. p. 337–80.
17. Birks RI, Huxley HE, Katz B. The fine structure of the neuromuscular junction of the frog. J Physiol. 1960;150:134–44.
18. Heuser JE, Reese TS, Landis DM. Functional changes in frog neuromuscular junctions studied with freeze-fracture. J Neurocytol. 1974;3:109–31.
19. Sato I, Shimada K, Ezure H, Omata H, Kawagoe T, Sato T. Morphology of the nerve terminals of laryngeal muscles in the Japanese Monkey (Macaca fuscata). Okajimas Folia Anat Jpn. 1995;72:1–6.
20. Rossi G, Cortesina G. Morphological study of the laryngeal muscles in man. Insertions and courses of the muscle fibres, motor end-plates and proprioceptors. Acta Otolaryngol. 1965;59:575–92.
21. Sonesson B. On the anatomy and vibratory pattern of the human vocal folds. Acta Otolaryngol. 1960;57:320.
22. Yoshihara T, Kanda T, Yaku Y, Kaneko T. Neuromuscular junctions of the posterior cricoarytenoid muscle in the human adult, human fetus and cat. Histochemical and electron microscopic study. Acta Otolaryngol. 1984;97:161–8.
23. Katto Y. Muscle spindle in intrinsic muscles of human larynx: distribution and morphological features. Nihon Jibiinkoka Gakkai Kaiho. 1987;90:59–70.
24. Katto Y. Morphological observation on silver-impregnated nerve endings in human interarytenoid muscle. J Jpn Bronchoesophagol Soc. 1990;41:270–6.
25. Katto Y, Okamura H, Yanagihara N. Observation on nerve endings in human intrinsic laryngeal muscles by using silver impregnation method. J Jpn Bronchoesophagol Soc. 1987;38:409–14.
26. Gambino DR, Malmgren LT, Gacek RR. Three-dimensional computer reconstruction of the neuromuscular junction distribution in the human posterior cricoarytenoid muscle. Laryngoscope. 1985;95:556–60.
27. Rosen M, Malmgren LT, Gacek RR. Three-dimensional computer reconstruction of the distribution of neuromuscular junctions in the thyroarytenoid muscle. Ann Otol Rhinol Laryngol. 1983;92:424–9.
28. Freije J, Malmgren LT, Gacek RR. Motor endplate distribution in the human lateral cricoarytenoid muscle. Arch Otolaryngol Head Neck Surg. 1986;112:176–9.
29. Freije J, Malmgren LT, Gacek RR. Motor endplate distribution in the human interarytenoid muscle. Arch Otolaryngol Head Neck Surg. 1987;113:63–8.
30. De Vito MA, Malmgren LT, Gacek RR. Three dimensional distribution of neuromuscular junctions in human cricothyroid. Arch Otolaryngol. 1985;111:110–3.
31. Hisa Y, Tadaki N, Uno T, Koike S, Tanaka M, Okamura H, Ibata Y. Nitrergic innervation of the rat larynx measured by nitric oxide synthase immunohistochemistry and NADPH-diaphorase histochemistry. Ann Otol Rhinol Laryngol. 1996;105:550–4.
32. Hisa Y, Uno T, Tadaki N, Murakami Y, Okamura H, Ibata Y. Distribution of calcitonin gene related peptide nerve fibers in the canine larynx. Eur Arch Otorhinolaryngol. 1992;249:52–5.
33. Neuhuber WL, Worl J, Berthoud HR, Conte B. NADPH-diaphorase-positive nerve fibers associated with motor endplates in the rat esophagus: new evidence for co-innervation of striated muscle by enteric neurons. Cell Tissue Res. 1994;276:23–30.
34. Hisa Y, Tadaki N, Uno T, Okamura H, Taguchi J, Ibata Y. Calcitonin gene-related peptide-like immunoreactive motor neurons innervating the canine inferior pharyngeal constrictor muscle. Acta Otolaryngol. 1994;114:560–4.
35. Takami K, Kawai Y, Uchida S, Tohyama M, Shiotani Y, Yoshida H, Emson PC, Girgis S, Hillyard CJ, MacIntyre I. Effect of calcitonin gene-related peptide on contraction of striated muscle in the mouse. Neurosci Lett. 1985;60:227–30.
36. New HV, Mudge AW. Calcitonin gene-related peptide regulates muscle acetylcholine receptor synthesis. Nature. 1986;323:809–11.
37. Yoshihara T, Nomoto M, Kanda T, Ishii T. Ultrastructural and histochemical study of the neuromuscular junctions in the denervated intrinsic laryngeal muscle of the cat. Acta Otolaryngol. 1991;11:607–14.

Autonomic Nervous System

Hideki Bando, Ken-ichiro Toyoda, and Yasuo Hisa

H. Bando • K. Toyoda • Y. Hisa (✉)
Department of Otolaryngology-Head and Neck Surgery, Kyoto Prefectural University of Medicine,
Kawaramachi-Hirokoji, Kamigyo-ku, Kyoto 602-8566, Japan
e-mail: yhisa@koto.kpu-m.ac.jp

Y. Hisa (ed.), *Neuroanatomy and Neurophysiology of the Larynx*, DOI 10.1007/978-4-431-55750-0_4

4.1 Introduction

Autonomic nerves control involuntary and reflective activity in various organs and tissues such as the heart, respiratory system, digestive tract, blood vessels, glands, and smooth muscles, which contributes to the maintenance of life activities.

Autonomic nervous system is conventionally divided into two subtypes, sympathetic and parasympathetic nervous systems, and is defined as the peripheral efferent fibers. However, recent studies have revealed the existence of afferent fibers (general visceral afferent fiber).

The autonomic innervation of the larynx had been veiled; however, recent studies including ours have identified the autonomic nerve fibers in the larynx and illuminated the distribution of autonomic innervation.

4.2 Nerve Ending

The anatomical distribution of ganglions and postganglionic neurons is different between sympathetic and parasympathetic nervous systems. While sympathetic postganglionic neurons are distributed systematically along the sympathetic trunk, distributions of parasympathetic postganglionic neurons are limited to inside or vicinities of target organs. Sympathetic and parasympathetic nervous systems play contrastive roles in controlling of physiological functions of the target organs.

Preganglionic fibers of both sympathetic and parasympathetic nervous systems are composed of moderately myelinated group B nerve fiber with high conduction velocity. The neurotransmitter of both fibers is acetylcholine (ACh). On the other hand, postganglionic nerves are composed of group C fiber with low conduction velocity. The neurotransmitter of the nerve ending of sympathetic nerves is noradrenaline (NA), while Ach is for parasympathetic nerves and certain sympathetic nerves.

The stimulation to preganglionic fibers is also transmitted by non-noradrenergic non-cholinergic (NANC) nerves beside these two main transmitters. The transmitters for NANC include purine bodies (ATP), neuropeptide (NPY, VIP, etc.), and gaseous transmitters (NO). These transmitters are co-localized or co-released with NA and ACh as the neurotransmitters or neuromodulators.

Identification of transmitters is quite important for morphological study of ANS. Falck-Hillarp method has been employed for optical microscopic identification of sympathetic noradrenergic innervation [1]. In the electron microscopic study, 5-hydroxydopamine (5-OHDA) has been used to identify the nerve endings [2]. Since immunohistochemistry had been developed, immunohistochemical detection of tyrosine hydroxylase (TH), enzyme for catecholamine synthesis, came to be employed for the study of noradrenergic neurons. Immunohistochemistry for TH is more reliable than conventional fluorescent histochemical technique, and permanent preservation of the specimen is possible. Moreover, TH immunohistochemistry is favorable for electron microscopic study and counter staining for revealing the precise localization.

On the other hand, detection of ACh in synaptic vesicles seems to be the most reliable method for observation of parasympathetic nerves. However, ACh is easily dissolved by hydrolysis and the chemical property makes detection in tissue specimen very difficult. Thus, immunohistochemistry for choline acetyltransferase (ChAT), the transferase enzyme responsible for synthesis of ACh, has been widely employed for detection of cholinergic nerves.

4.3 Autonomic Innervation in the Larynx

4.3.1 Sympathetic Nervous System

The postganglionic neurons of laryngeal sympathetic nervous system have their cell body mainly in the superior cervical ganglion. The preganglionic neurons originate in the gray matter of the upper thoracic spinal cord. The nerve fiber leads to the superior cervical ganglion via the ventral root, white ramus, and cervical sympathetic trunk. Sympathetic innervation of the larynx had been considered to be innervated along with superior or inferior laryngeal arteries and veins. However, our study using Falck-Hillarp method revealed that sympathetic innervation of the canine larynx is distributed via the superior laryngeal nerve and inferior laryngeal nerve [3].

We also tried to clarify the distribution of sympathetic innervation in the larynx since it had been veiled until then [4]. Our study using Falck-Hillarp method and TH immunohistochemistry revealed detailed distribution of sympathetic nerve fibers in the laryngeal arteries and glands in the supraglottis, glottis, and subglottis. Tanaka et al. describes that TH-immunoreactive nerve fibers were located in the vicinity of the basal lamina, but they never terminated or penetrated the basal lamina [5].

4.3.1.1 Distribution of Laryngeal Adrenergic Neurons

The distribution of adrenergic neurons in the canine larynx was studied by electron microscopy with 5-OHDA method [6]. The existence of the adrenergic terminals of the canine laryngeal glands has been revealed by electron microscopy and fluorescence histochemistry. Adrenergic fibers with fluorescent varicosities were observed around the base of the acini in the submucosa (◘ Fig. 4.1) and around the vessels in the intrinsic laryngeal muscles (◘ Fig. 4.2). In dogs treated with 5-OHDA, adrenergic terminals were found near the blood vessels, gland cells, and myoepithelial cells in the submucosal gland region.

To determine the origin and course of noradrenergic nerve fibers contained in laryngeal nerve, the larynx of the dog was investigated using fluorescence histochemistry [3]. Denervation of the superior and inferior laryngeal nerves was also performed following preservation of the superior and inferior laryngeal artery and vein. Then, the larynx was removed and treated with Falck-Hillarp method

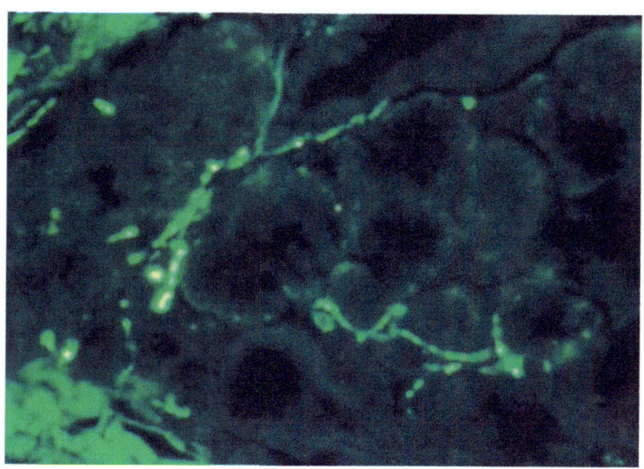

Fig. 4.1 Catecholamine-positive nerve fibers are observed around canine laryngeal glands (Falck-Hillarp method)

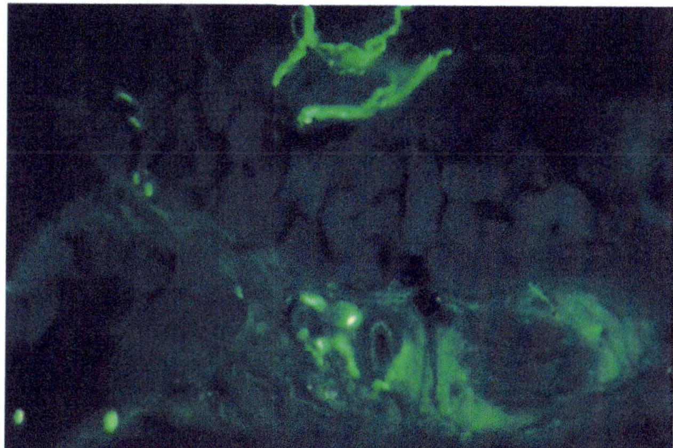

Fig. 4.2 Catecholamine-positive fibers are detected around the blood vessels of canine intrinsic laryngeal muscle (Falck-Hillarp method)

Table 4.1 The number of NA fibers in the denervated intrinsic laryngeal muscles

Denervated nerve	CT	TA	PCA	LCA	IA
SLN (E and I) ILN	–	–	–	–	–
SLN (I) ILN	→	–	–	–	–
SLN (E)	–	→	→	→	→
SLN (I)	→	↓	↓	↓	↓
ILN	→	↓	↓	↓	↓

Table 4.2 The number of NA fibers in the denervated laryngeal glands

Denervated nerve	Glottis	Supraglottis	Subglottis
SLN (E, I) ILN	–	–	–
SLN (E) ILN	–	–	–
SLN (E)	+	+	+
SLN (I)	–	–	+
ILN	+	+	y

(Tables 4.1 and 4.2). The superior laryngeal nerve and the recurrent laryngeal nerve were found to contain many noradrenergic nerve fibers. The supraglottic and subglottic submucosal glands received the adrenergic nerve fibers from the internal branch of the superior laryngeal nerve and the recurrent laryngeal nerve, respectively. The cricothyroid muscle received noradrenergic nerve fibers from the external branch of the superior laryngeal nerve, and other intrinsic muscles received them from the internal branch of the superior laryngeal nerve and the recurrent laryngeal nerve. The noradrenergic nerve fibers in the superior laryngeal nerve were originated from the superior cervical ganglion. The recurrent laryngeal nerve contains noradrenergic nerve fibers originating from the middle cervical ganglion and the superior cervical ganglion via the vagal nerve. The denervation study revealed that these laryngeal nerves were the only noradrenergic nerve supply for the larynx.

Immunohistochemical study of TH in the canine larynx also supports the result of these studies [7]. Many tyrosine hydroxylase-immunoreactive nerve fibers were observed around arteries in the laryngeal mucosa and intrinsic laryngeal muscles (Fig. 4.3a). However, TH-immunoreactive nerve fibers were rarely expressed around the capillaries. Thus, we concluded that the capillaries in the laryngeal mucosa are not under the control of noradrenergic neurons. The glottis and the supraglottic mucosa contained large number of TH-positive nerve fibers, while there were only small numbers of positive fibers in the subglottic mucosa. In the glandular region, some of the TH-immunoreactive fibers terminated around the basement membranes of the glandular cells (Fig. 4.3b). These findings are compatible with the result of 5-OHDA method and suggest that TH-immunoreactive fibers may be directly concerned in the control of blood flow and the regulation of glandular secretion.

4.3.2 Parasympathetic Nervous System

Parasympathetic nervous system plays major roles in the motor control of mucus secretion in the larynx. The cell body of postganglionic neuron is considered to present in the intralaryngeal ganglion [8–10]. As Yoshida et al. describe, intralaryngeal ganglionic neurons have cholinergic nature and innervate vessels and glands [9]. The cell body of preganglionic neuron is located in the dorsal nucleus of vagal nerve and projects the nerve fiber to the larynx via the vagal nerve and superior or inferior laryngeal nerve.

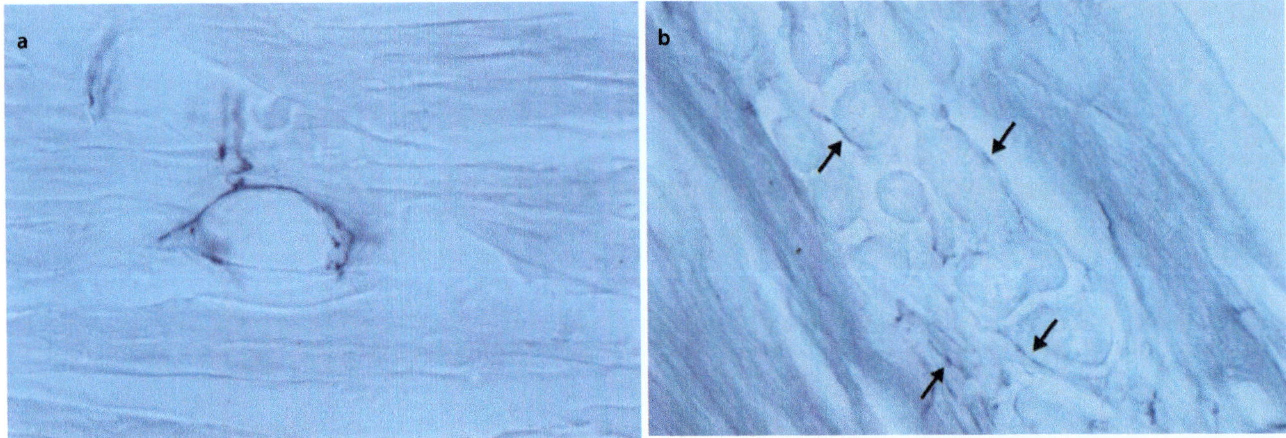

Fig. 4.3 Immunohistochemistry for TH in the canine intrinsic laryngeal muscle. (**a**) TH-positive nerve fibers are observed around the arteries. (**b**) TH-positive nerve fibers are observed around the laryngeal glands

ACh is known as the transmitter of motor neurons, pre- and postganglionic nerve fibers of parasympathetic nerves and some of the sympathetic nerves. Recently, ACh is also reported to work as the transmitter of a part of sensory nervous system [11, 12].

As it is difficult to fix ACh in the tissue because of its chemical structure, histochemical study on acetylcholinesterase (AChE), the hydrolase for acetylcholine, had been employed for detection of peripheral cholinergic neurons formerly. However, reliability of this method is not certain, as AChE is distributed also in non-cholinergic neurons. Moreover, motor neurons of intrinsic laryngeal muscles have not been identified by AChE histochemistry though they are considered to be cholinergic.

On the other hand, choline acetyltransferase (ChAT) has been targeted in immunohistochemical study of central cholinergic nerves. However, peripheral cholinergic nerves are not identified by conventional anti-ChAT antibody. Peripheral ChAT (pChAT), splicing variant of conventional central ChAT, was recently been identified, and distribution of peripheral cholinergic nerves in various organs has been illuminated [13–16]. Immunohistochemistry for pChAT has allowed the detection of peripheral cholinergic neurons (Fig. 4.4).

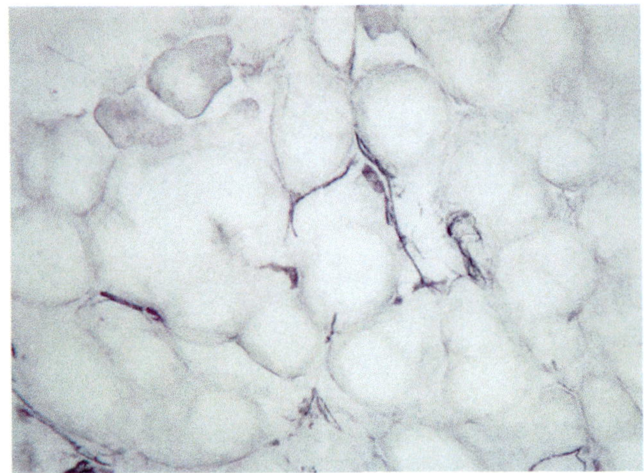

Fig. 4.4 Immunohistochemistry for pChAT. The pChAT immunoreactivity is observed in the nerve fibers around rat laryngeal glands

Nakanishi et al. described the study of parasympathetic innervation of rat larynx using pChAT immunohistochemistry. pChAT-positive nerve fibers were distributed laryngeal glands (gaps between acinar cells and periphery of glands), mucosa, blood vessels, and intrinsic laryngeal muscles. Double staining for pChAT immunohistochemistry and AChE histochemistry proved that all the pChAT-positive fibers showed AChE activity. However, there were a few AChE-positive fibers which did not show pChAT immunoreactivity.

4.3.2.1 New Splicing Variant of Choline Acetyltransferase (cChAT-v)

In the process for cloning of pChAT, another splicing variant of ChAT was detected, and analysis of the DNA structure revealed that it lacks exons 7 and 8 of ChAT gene. The new variant was named cChAT-v as the second splicing variant (Fig. 4.5).

To illuminate the expression of cChAT-v in the larynx, detection of ChAT mRNA in rat laryngeal tissue was performed as the first step. Rat laryngeal mRNA was harvested and applied for RT-PCR (Fig. 4.6). The expression of cChAT-v and pChAT was confirmed while no expression of cChAT was observed.

We conducted immunohistochemical study to illuminate the distribution of ChAT-positive nerve fibers (cChAT, pChAT, and cChAT-v) in rat laryngeal innervation. The primary antibody for cChAT-v was synthesized from 22 amino acids equivalent to the junction between exon 6 and exon 9 of rat ChAT gene.

1. Nucleus Ambiguus
   Some nucleuses of cChAT-positive motor neurons were detected in the nucleus ambiguus, and cChAT-v positive cells were also distributed. However, pChAT was not expressed in the nucleus ambiguus (Fig. 4.7).

2. Dorsal nucleus of vagal nerve
   Immunoreactivities of cChAT and cChAT-v were confirmed. pChAT was not observed in the dorsal nucleus. However, pChAT immunoreactivity was confirmed 7 days after unilateral vagotomy (◘ Fig. 4.8).
3. Intrinsic laryngeal muscles
   cChAT and cChAT-v immunoreactivities were confirmed in the nerves connected to neuromuscular junctions in the middle portion of the muscles. A small number of pChAT-positive nerves were also observed (◘ Fig. 4.7).
4. Laryngeal glands
   pChAT immunoreactivities were confirmed in the nerves around the acinar cells of laryngeal glands. cChAT-v was detected only in the large fibers. cChAT was not observed in the laryngeal glands (◘ Fig. 4.9).
5. Submucosal tissue
   Nerve fibers with pChAT and cChAT-v immunoreactivities were confirmed in the submucosa of supraglottic region, while cChAT was not observed (◘ Fig. 4.5).
6. Intralaryngeal ganglion
   Cell bodies of intralaryngeal ganglion were stained by pChAT and cChAT-v immunohistochemistry. cChAT immunoreactivity was confirmed only in the nerve fibers around the ganglion (◘ Fig. 4.9).
7. Nodose ganglion
   pChAT- and cChAT-v-positive nerves were observed. cChAT was not expressed in nodose ganglion. These results indicate that pChAT and cChAT-v are expressed not only in parasympathetic nerves but in sensory nervous systems (◘ Fig. 4.10).

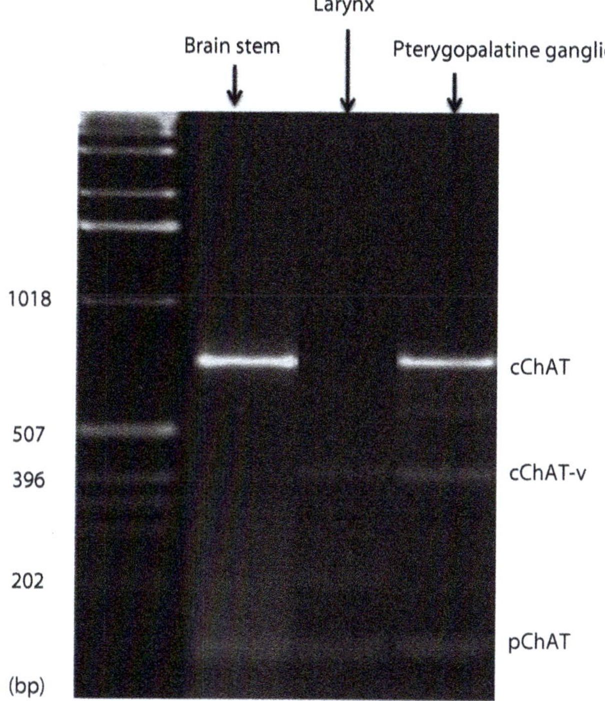

◘ **Fig. 4.5** Structure of ChAT variants

Although the expression of cChAT-v in the laryngeal nervous system and the mucosal tissue has been revealed by our study, the contribution in the laryngeal parasympathetic innervation is still unclear. Further study is required to illuminate the roles of each subtype.

4.3.2.2 Expression of ChAT in the Vagotomized Laryngeal Tissue

Unilateral vagotomy was performed in order to examine the effect of parasympathetic innervation on the expression of cChAT-v. In the intrinsic laryngeal muscles, all of cChAT-, cChAT-v-, and pChAT-immunolabeled nerve fibers were disappeared, while no alteration was observed in the intralaryngeal ganglion, laryngeal gland, and the small vessels. In the dorsal motor neuron of denervated vagus, the number of cChAT- and cChAT-v-immunolabeled neuron was gradually decreased from the second day. On the other hand,

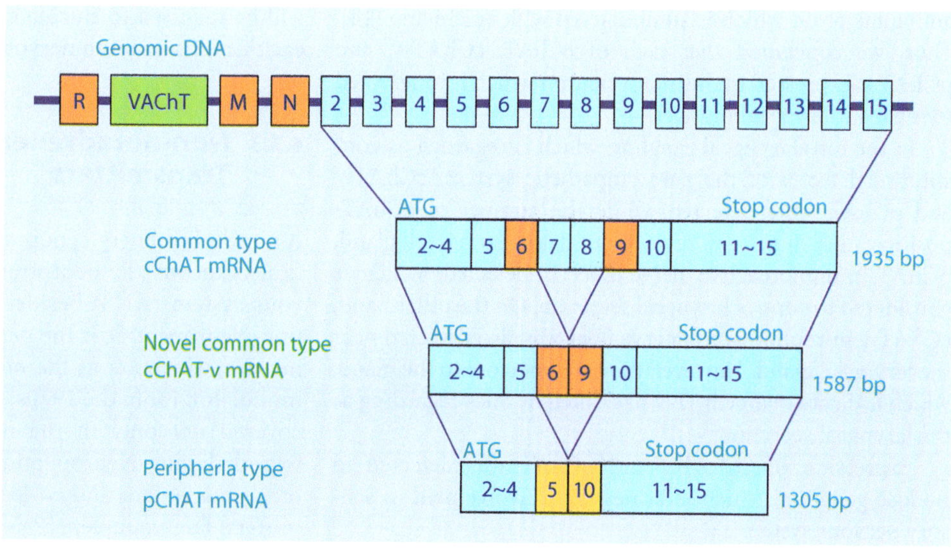

◘ **Fig. 4.6** The expression of ChAT mRNA in the brain stem, larynx, and pterygopalatine ganglion. The expression of cChAT-v and pChAT was confirmed in the RT-PCR of rat larynx. (cChAT 755bp, cChAT-v 407bp, pChAT 125bp)

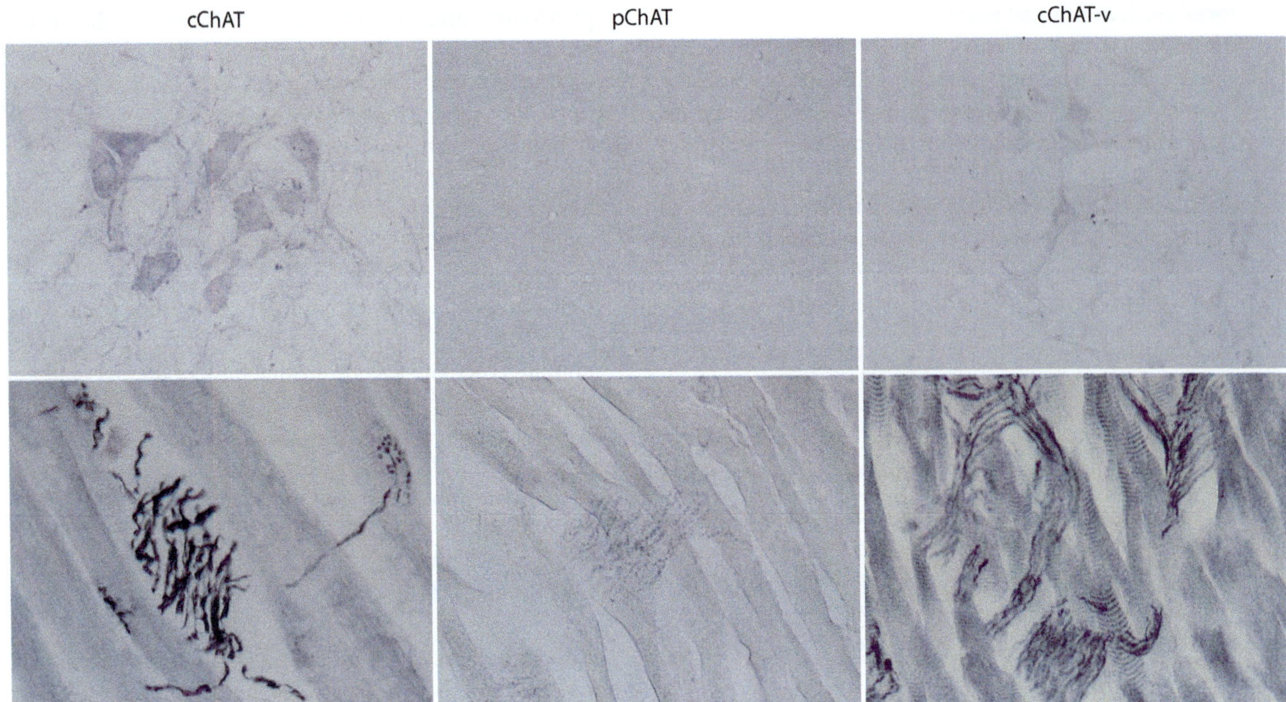

Fig. 4.7 Distribution of ChAT variants in laryngeal motor neurons. Neurons immunolabeled with cChAT and cChAT-v are observed in the nucleus ambiguus, while pChAT-immunolabeled neuron is absent. However, nerve fibers in the intrinsic laryngeal muscles show immunoreactivity for each of ChAT variants including pChAT

pChAT-immunolabeled neuron started to appear at the fourth day and increased at the seventh day.

The result of RT-PCR study shows constitutional expression of cChAT-v and pChAT in the larynx, and immunohistochemical study revealed the detailed expression of each subtypes of ChAT in the laryngeal nervous system including the brain stem.

Expression of cChAT and cChAT-v is observed in the motor neurons in the nucleus ambiguus and nerve fibers in the intrinsic laryngeal muscles. Though pChAT was not detected in the nucleus ambiguus, Nakanishi described that pChAT-immunoreactive neuron is detectable in the nucleus ambiguus of rat which is administered with colchicine [14]. Thus, we concluded that each of cChAT, cChAT-v, and pChAT is expressed in the motor neuron which is innervated to intrinsic laryngeal muscles.

In the intralaryngeal ganglion which is regarded as constitutional factor of the parasympathetic system, cChAT-v and pChAT were detected, while the number of cChAT-positive fiber was very few. Around the laryngeal gland, pChAT-immunoreactive nerve fiber is detected, which is considered to control laryngeal secretion. On the other hand, cChAT-v immunoreactive nerve fiber was also detected near the laryngeal gland. However, they are not close to the gland, which indicate that cChAT-v has different roles regarding as the laryngeal secretion.

Expression of cChAT-v and pChAT was detected in nodose ganglion, which indicates their involvement in sensory nervous system.

Expression of cChAT-v and pChAT in intralaryngeal ganglion and nerve fibers around the laryngeal glands was not altered after cervical vagotomy. This result shows that laryngeal glands are innervated by postganglionic fibers of intralaryngeal ganglion as previously reported [10, 17, 18], and these fibers are considered to be cChAT-v- and pChAT-positive cholinergic nerves.

In the brain stem, cChAT expression in the dorsal nucleus of the vagus was apparently decreased after vagotomy, and expression of pChAT was rather increased (Fig. 4.8). Previous studies describe that expression of cChAT was recovered after a long period elapsed [19–21]. Further study will be required to elucidate the pathophysiological roles of each ChAT variant in nervous disorders.

4.3.3 Non-noradrenergic, Non-cholinergic Transmitters

A non-noradrenergic, non-cholinergic transmitter (NANC) is also known as a neurotransmitter of the autonomic nervous system (ANS) besides noradrenaline and acetylcholine. Neuropeptide is the peptide produced and secreted by neuron, which act as the neurotransmitter and the neuromodulator. From the 1970s, various peptides have been discovered not only in the neurosecretory neurons in the hypothalamus. Neuropeptides are localized in the neurons of various organs and co-localized with classic neurotransmitters. For example, NPY is co-localized with noradrena-

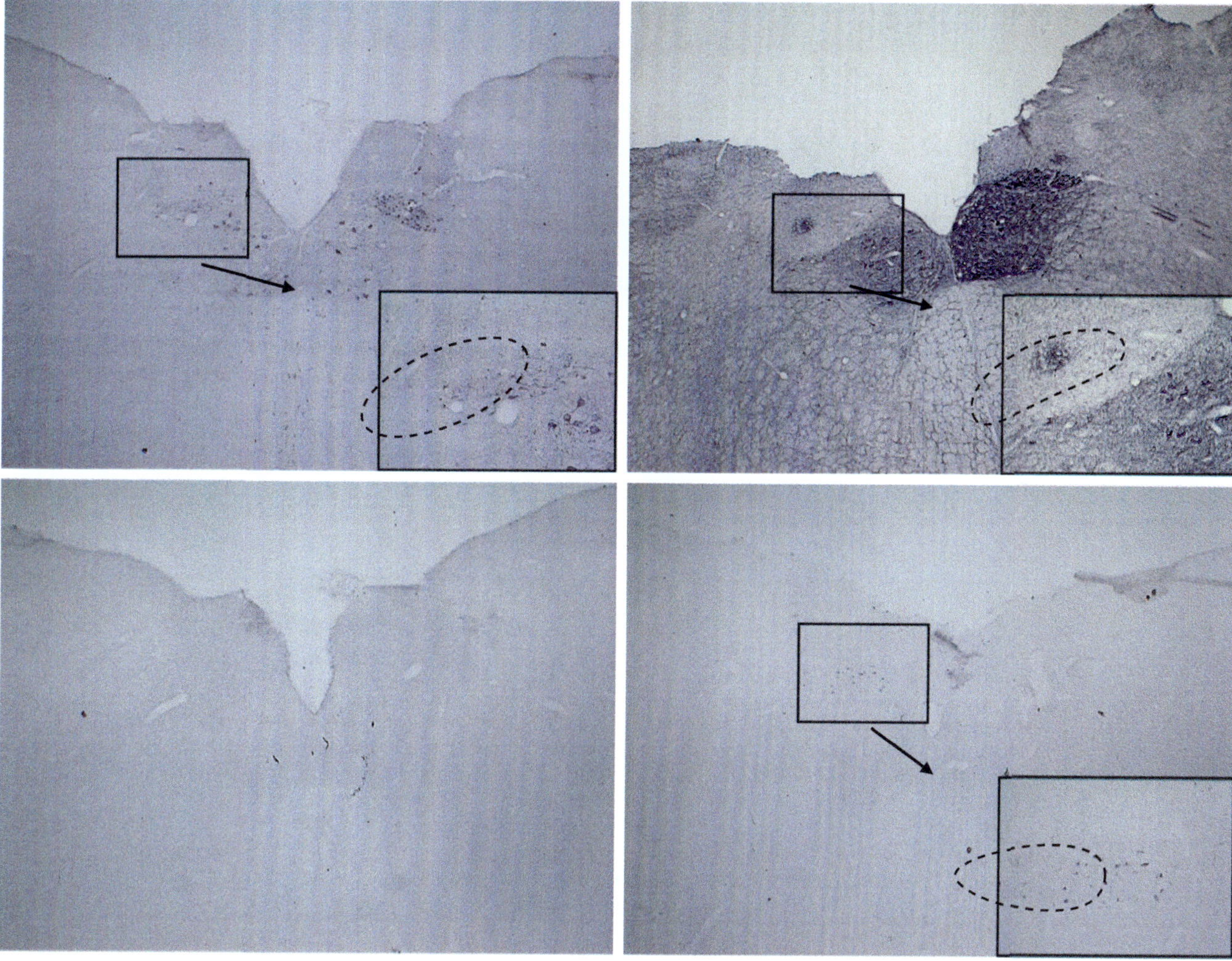

Fig. 4.8 Chronological changes of expression of ChAT in the dorsal nucleus of vagal nerve after unilateral vagotomy. 10a, b:cChAT, 10c,d: pChAT 10a. (Day 2) cChAT expression in the denervated side is decreased compared to the contralateral side. 10b (Day 7) The expression in the denervated side showed further decrease. 10c (Day 2) pChAT expression was not observed in both sides. 10d (Day 7) pChAT expression was confirmed in the denervated side

line and ATP in the peripheral sympathetic nervous system, and VIP is co-released with acetylcholine in the postganglionic nerve fibers of the parasympathetic nervous system. Recent studies revealed the existence of neuropeptides in the laryngeal nervous system, and their roles have been clarified gradually.

The following are representative neuropeptides in the larynx:

- NPY: Neuropeptide Y (NPY) is a 36-amino-acid neuropeptide which was originally isolated from porcine brain, and the molecular structure was revealed in 1982 by Tatemoto [22]. NPY is co-localized with noradrenaline in the nerve endings of sympathetic nerves and released together. NPY, together with the sympathetic nerves, may be important in angiogenesis during tissue development and repair [23].
- VIP: Vasoactive intestinal peptide (VIP) is a 28-amino-acid peptide isolated from porcine intestine in 1974. VIP is widely distributed over both central and peripheral nervous systems. VIP is important as a regulator for controlling the relaxation of intestinal smooth muscles, dilatation of peripheral blood vessels, and salivary secretion [24].
- CGRP: Calcitonin gene-related peptide (CGRP), a splicing variant of the calcitonin gene, is a 37-amino-acid neuropeptide which was isolated in the 1980s. CGRP is known to be distributed widely over both central and peripheral nervous systems. The principal roles of CGRP are dilatation of vessels and transmission of pain and inflammation [25].

In 1995, Tanaka et al. reported detailed distribution of VIP, NPY, and TH in the larynx using immunohistochemical study with electron microscopy [5]. VIP-positive fibers were located around the basal membrane and myoepithelial cells of laryngeal glands. Some of these fibers contacted with the basal lamina, and some of them pierced it and ran intercellularly in the adjoining glandular cells without making synaptic contacts with them.

NPY- and TH-positive fibers were located in the vicinity of the basal lamina, but they were less abundant than VIP-positive fibers at this region. They never terminated or pen-

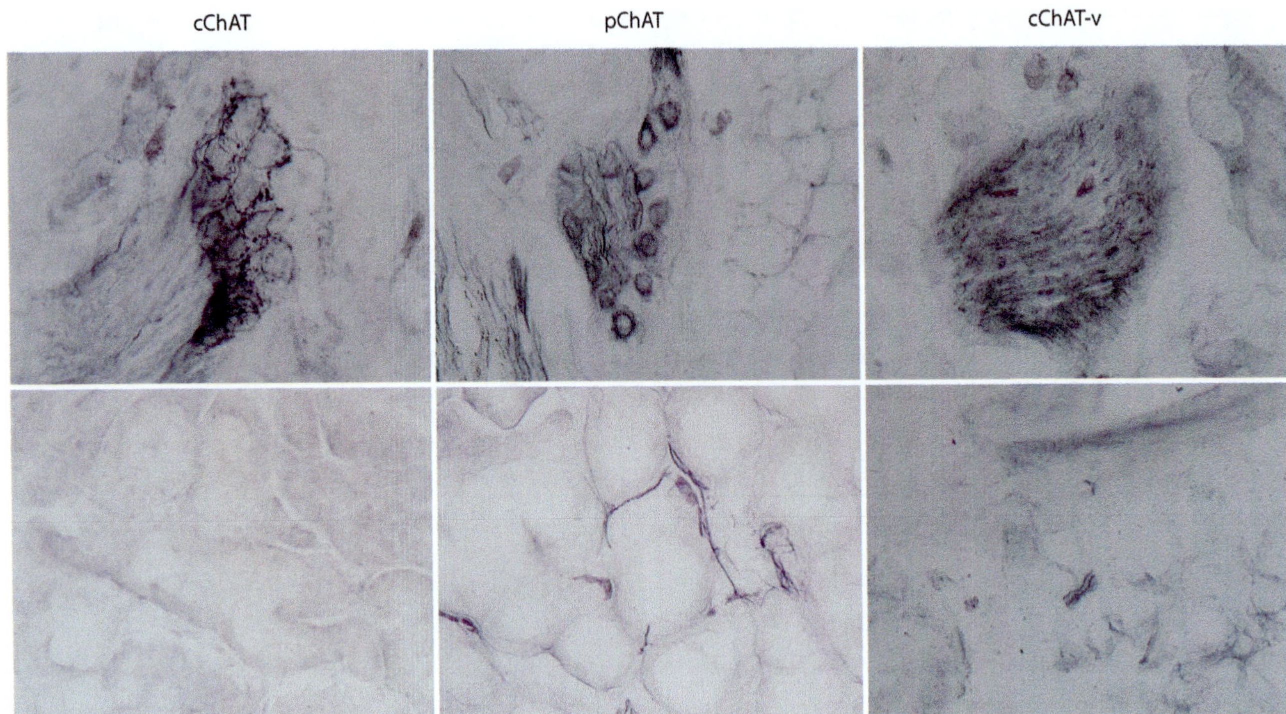

Fig. 4.9 The ChAT variants expression in laryngeal parasympathetic system. Immunolabeled neurons with pChAT and cChAT-v are observed in the intralaryngeal ganglion, while cChAT is observed only in the fibers around the cell body. In the laryngeal gland, pChAT is observed around the acini, while no expression of cChAT was observed. The cChAT-v immunoreactivity was limited to the larger nerves

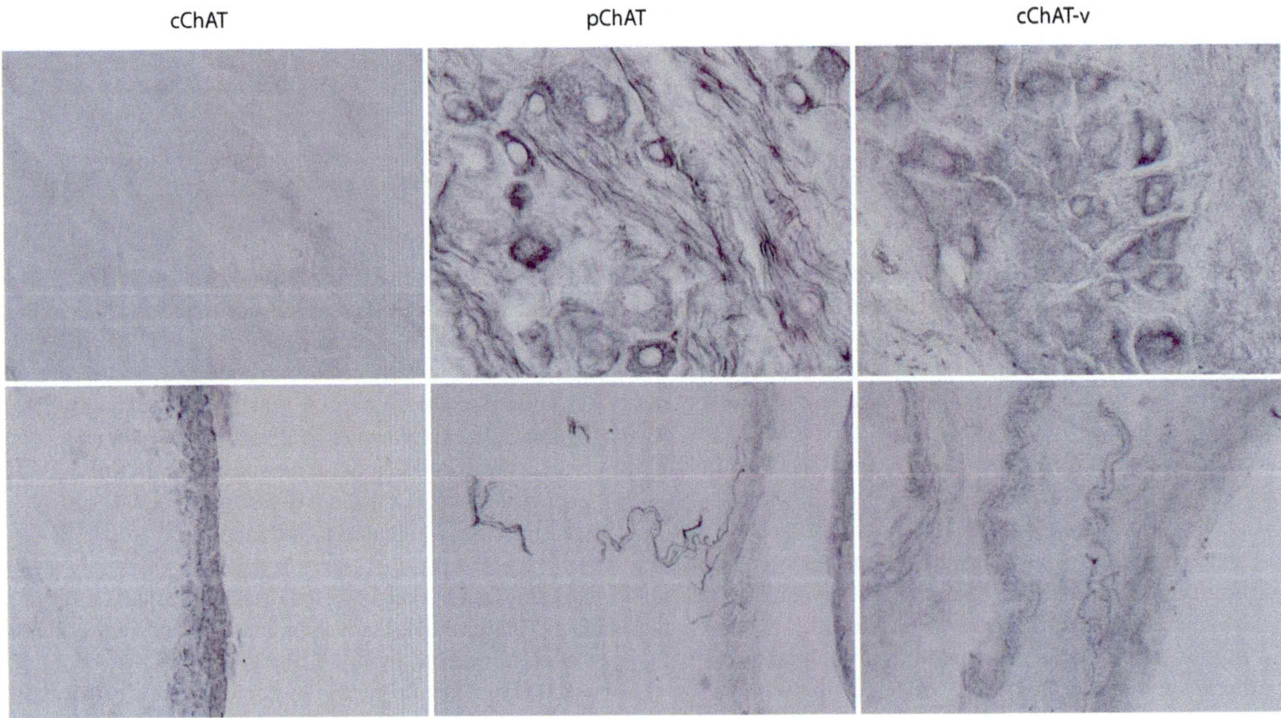

Fig. 4.10 Expression of ChAT variants in the laryngeal sensory system. The cell body immunolabeled with pChAT and cChAT-v is detected, while cChAT is not observed

etrated the basal lamina. The pattern of distribution of TH-positive fibers was similar to that of NPY-positive fibers. The estimated ratio of VIP-, TH-, and NPY-positive fibers in the laryngeal glands was 20:4:1 from the density of fibers. This value was equal between serous and mucous glandular cells, which indicates that both may receive the same pattern of autonomic innervation.

Domeij et al. describe that fibers with NPY and VIP immunoreactivity coexisted in local AChE-positive ganglionic cells and in a subpopulation of the AChE-positive fibers [26]. The result of vagotomy and removal of the SCG suggests that VIP is present in the postganglionic parasympathetic innervation, whereas NPY is present in both the postganglionic parasympathetic and sympathetic innervations of the rat larynx.

Kawasoe et al. reported that VIP, SP, and CGRP were expressed around the arteries in the lamina propria of the canine larynx [27].

4.3.4 Distribution of CGRP-Positive Nerve Fibers in the Larynx

The distribution pattern of calcitonin gene-related peptide (CGRP) nerve fibers was investigated by immunohistochemistry in the laryngeal mucosa, glands, and intrinsic laryngeal muscles of the dog. CGRP-immunoreactive nerve fibers were found more frequently than substance P-immunoreactive nerve fibers in every region of the larynx [28]. CGRP-immunoreactive nerve fibers were mainly found in the epithelium of the epiglottis, arytenoid region, and subglottis. In the lamina propria, the plexus of CGRP nerve fibers was present, with some of these fibers associated with blood vessels. Laryngeal glands were also innervated by a few CGRP nerve fibers. In the intrinsic laryngeal muscles, abundant immunoreactivity was observed, and many motor end plate-like structures were found with CGRP immunoreactivity. These findings strongly suggest that CGRP plays an important role in all of the sensory, motor, and autonomic nervous systems of the larynx.

Regarding the gasotransmitters in the canine larynx, our study revealed that NADPH diaphorase (NADPHd) known as a nitric oxide synthase is expressed in the vicinity of acinar cells of laryngeal glands and vessels [29]. Some of them were coexisted with VIP or calcitonin gene-related peptide (CGRP) [29] (Figs. 4.11 and 4.12). In the intralaryngeal ganglion, while NADPHd was coexisted with VIP in the cell bodies, no cell body expressed both NADPHd and CGRP. These results support a hypothesis that NO is concerned in the regulation of laryngeal gland secretion in corporate with VIP existing in the larynx or CGRP transported to the larynx.

4.4 Receptors of Autonomic Nervous System

Autonomic receptors are divided into adrenergic receptors and cholinergic receptors that respond to the binding of catecholamine and acetylcholine, respectively. In the nerve endings of sympathetic and parasympathetic preganglionic nerve fibers, nicotinic acetylcholine receptors are distributed. On the other hand, the adrenergic receptor and the muscarinic cholinergic receptor are located in various organs as the target of sympathetic nerves and parasympathetic nerves, respectively.

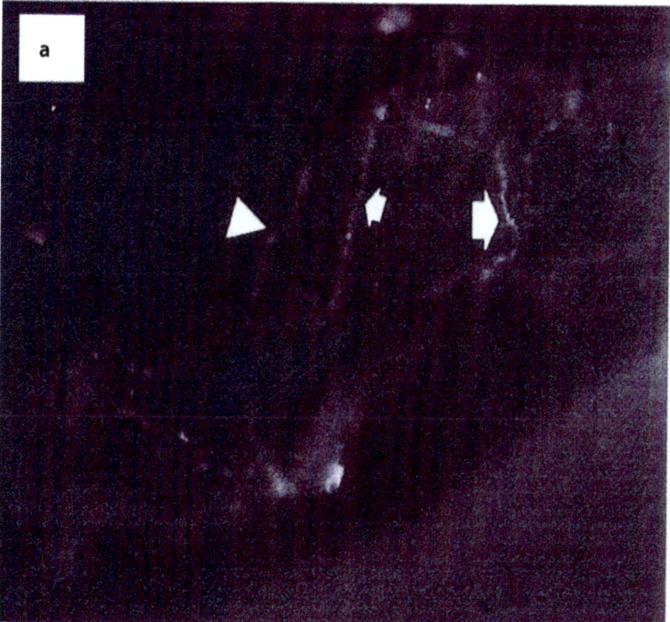

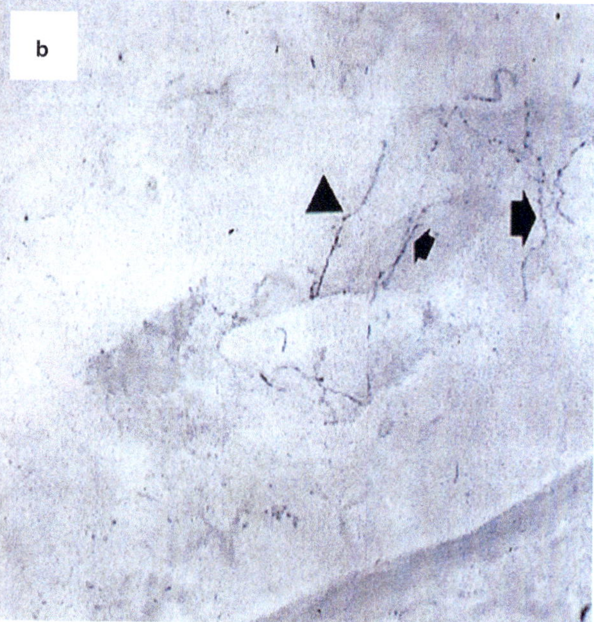

Fig. 4.11 Double staining with immunohistochemistry for VIP (**a**) and NADPHd histochemistry (**b**). Double-stained nerve fibers are distributed around the arteries in canine laryngeal mucosa (*arrows*)

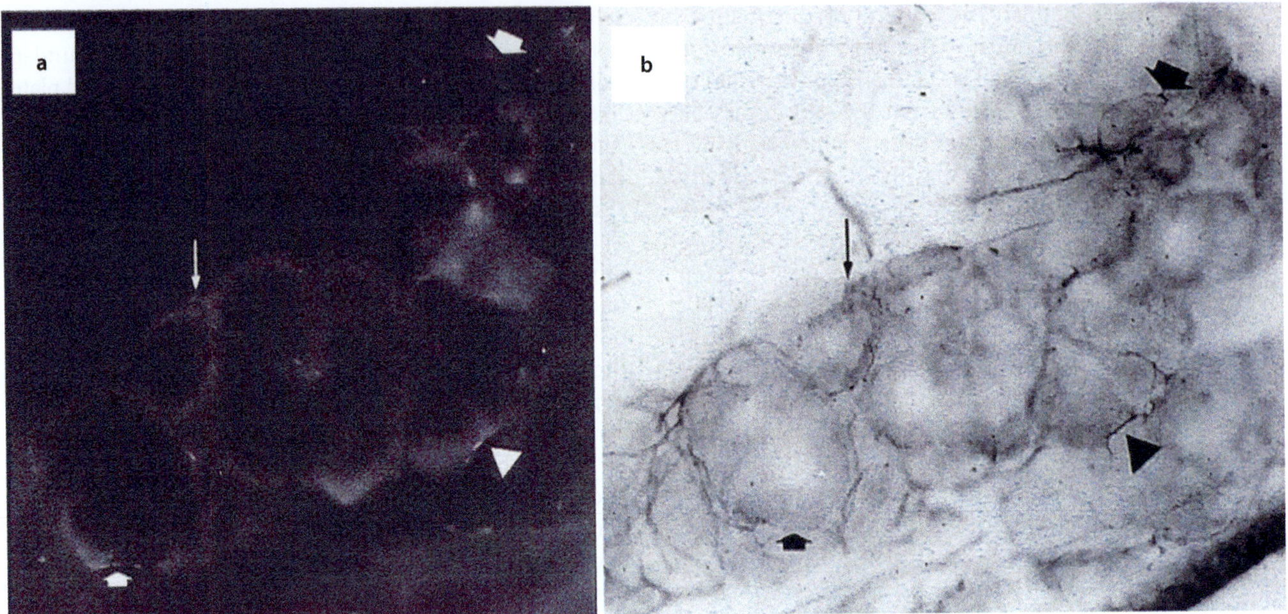

Fig. 4.12 Double staining with immunohistochemistry for CGRP (**a**) and NADPHd histochemistry (**b**). Double-stained nerve fibers are distributed around canine laryngeal glands

4.4.1 Adrenergic Receptors

The adrenergic receptors are a class of G protein-coupled receptors which are activated by noradrenaline and epinephrine. Adrenergic receptors are divided into two main types, **α** and β, which include several subtypes alpha1 and alpha2 and beta1, beta2, and beta3. Beta2-adrenergic receptor is known to be widely distributed in the respiratory tract, particularly in the airway smooth muscle, and involved in relaxation of airway smooth muscle in the airway [30]. However, functional role in the laryngeal physiology has not been well understood.

Adrenergic control of mucus secretion can be demonstrated in the airways of experimental animals but appears to be of minor importance in humans.

4.4.2 Cholinergic Receptors

4.4.2.1 Muscarinic Acetylcholine Receptors

The parasympathetic nerves provide the dominant autonomic control of airway glands. They release acetylcholine onto muscarinic receptors, causing mucous secretion. The release of acetylcholine from the parasympathetic nerves is controlled by muscarinic autoreceptors located on the nerves.

Muscarinic acetylcholine receptors are integral membrane proteins that form G protein-receptor complexes in the cell membranes with the roles as the receptor stimulated by acetylcholine released from postganglionic fibers in the parasympathetic nervous system. Five subtypes (M1–M5 receptors) have been identified, and each subtype is expressed in many central and peripheral tissues including the airway [31].

Parasympathetic nervous system plays major roles in the motor control of mucus secretion in the airways. Submucosal glands express both M1 and M3 muscarinic receptors [33], and M3 muscarinic receptor is known to mediate cholinergic mucus secretion. The M1 receptor is involved in the control of electrolyte and water secretion with M3 receptor [32].

On the other hand, M2 receptors are located on cholinergic nerves where they act as inhibitory autoreceptors to activate a negative feedback system, whereby acetylcholine released from the nerves activates the receptor to inhibit further acetylcholine release [33].

The roles and expressions of adrenergic and cholinergic receptors in the laryngeal nervous systems have not been illuminated so far. However, our study has revealed the expressions of M2 and M3 receptors in the laryngeal submucosal glands (DATA not shown).

4.4.2.2 Nicotinic Receptors of the Autonomic Postganglionic Neuron

Acetylcholine released from preganglionic nerves at the sympathetic and parasympathetic ganglion activates nicotinic receptors (nAChRs) of postganglionic neurons. nAChRs on autonomic neurons are typically composed of two α3 subunits in combination with three other AChR subunits. The study of transgenic mice showed that the α3 subunit is essential for autonomic ganglionic neurotransmission [34].

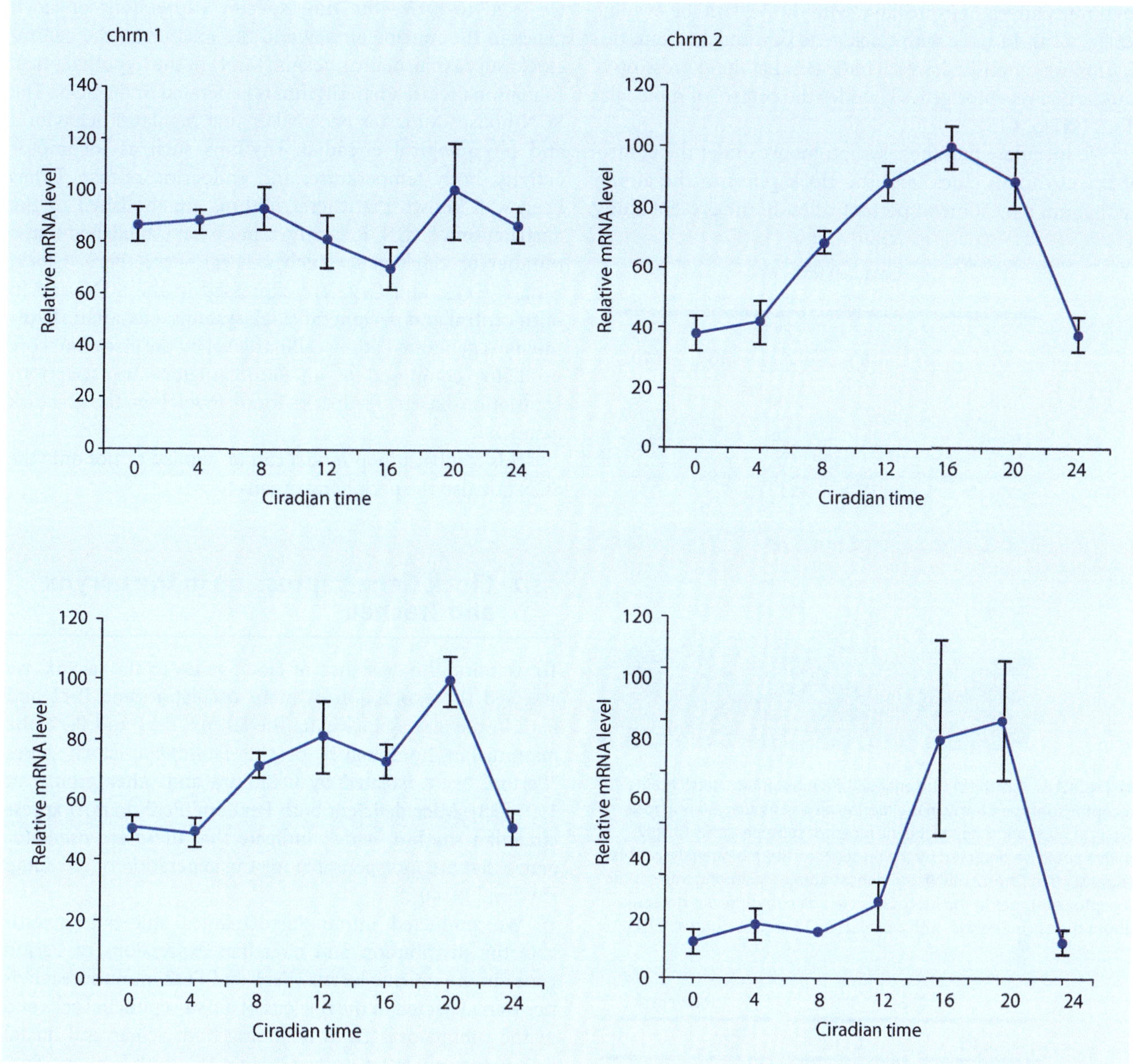

Fig. 4.13 The gene expression pattern of Chrm1, Chrm2, Chrm3, and Chrm4 by the real-time PCR. Expression of chrm2, 3, and 4 showed significant circadian rhythm peaked at CT12–20

4.5 Circadian Rhythm of the Laryngeal Parasympathetic Nervous System

Some symptoms of laryngeal and airway diseases are associated with specific times of the day. For example, nocturnal worsening of cough is a frequent complaint of patients with laryngitis and bronchitis. It is well known that there are diurnal cycles in respiratory physiology [35], and the circadian change of the parasympathetic activity is considered to be responsible for these rhythms. However, molecular mechanism for regulation of circadian change in motor control of parasympathetic innervation had not been clarified. We examined the circadian expression of mRNA of muscarinic receptors in the murine airway with real-time PCR and Northern blotting to reveal the molecular basis of circadian rhythm of the physiology of respiratory system [36, 37].

4.5.1 Circadian Expression of Muscarinic Receptors in the Murine Airway

The expression of mRNA of M2, M3, and M4 showed significant rhythmic expression, but no significant rhythm was observed in M1. In the former three genes, mRNA expression of M2 showed remarkable circadian rhythm peaked at CT12–20 and had a 2.6-fold amplitude with a peak at CT16 (Fig. 4.13). We speculated that M2 is a clock-controlled gene in the airway. Northern blot analysis showed a similar rhythmic pattern of expression, with the peak in the evening (Fig. 4.14). In mice with clock gene Cry knocked out, this rhythm was abolished, which indicates that the expression of muscarinic receptor genes ia under the control of molecular clock (Fig. 4.15).

We speculate that these symptoms are under the control of the circadian clock, and the clock genes in the airway epithelium play some important roles. In the present study,

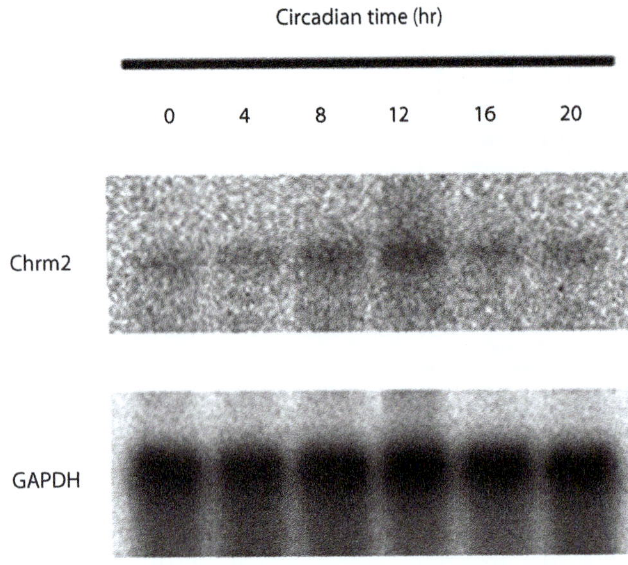

Fig. 4.14 Northern blot analysis of a muscarinic acetylcholine receptor subtype Chrm2 in the murine airway tissues. Among four types of muscarinic acetylcholine receptor subtypes, only Chrm2 mRNA could be detected by the present Northern blot analysis. This suggests that Chrm2 mRNA is the most abundant among muscarinic receptor subtypes in the lung. Clear circadian rhythm was detected about the Chrm2 mRNA with a peak at CT12 and a trough at CT0

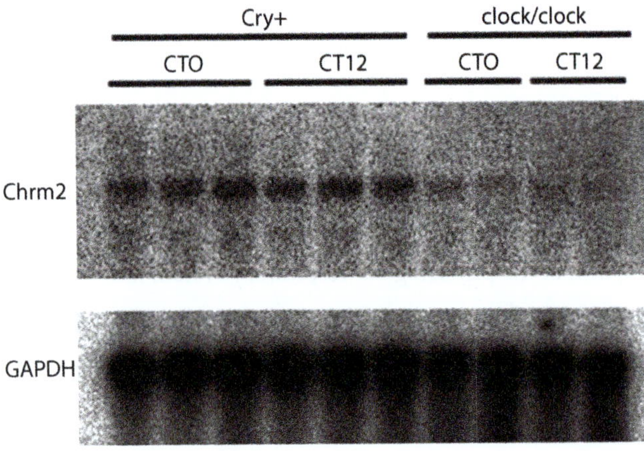

Fig. 4.15 Chrm2 mRNA in the lung of mCry1-/-mCry2-/- mice and Clock mutant mice. Both of them showed no circadian variation. mCry-double knockout mice showed high Chrm2 mRNA level in the lung, and clock mutant mice showed low mRNA level

we tried to prove the time-specific expressions of clock genes in the murine airway and the relation to the central clock: suprachiasmatic nucleus (SCN) in the hypothalamus. In mammals, circadian rhythm is generated in the SCN. The SCN houses a master pacemaker that regulates behavioral and physiological circadian rhythms such as locomotor activity, body temperature, and endocrine release. It has been well known that these rhythms are abolished by the destruction of SCN. Recently, it has been established that a number of clock genes such as Per1, Per2, Per3, Clock, Bmal1, Cry1, and Cry2 are expressed in the SCN, and in both central and peripheral clock systems, circadian rhythmicity is generated at the cellular level by the circadian core oscillator composed of an autoregulatory transcription-translation-based feedback loop involving these clock genes.

This feedback loop model can be applied to not only the SCN but also the peripheral organs.

4.5.2 Clock Gene Expression in the Larynx and Trachea

To examine the existence of clock genes in the larynx, we assessed the expression of main oscillator gene Per1 and Per2 in the airway including the larynx. Per1 and Per2, the mammalian homologues of the *Drosophila* clock genes "Period," were isolated by Okamura and other groups in 1997 [38]. Mice deficient both Per1 and Per2 do not express circadian rhythm, which indicate that these are oscillator genes that are indispensable for the generation of circadian rhythm [39, 40].

We conducted immunohistochemical studies to investigate the distribution and circadian expressions of Period genes in the murine larynx. Per1 and Per2 immunoreactivities were detected in the nucleus of airway epithelial cells and in the submucosal gland including both acinar and ductal cells in the subglottic region (Fig. 4.16). Both Per1 and Per2 immunoreactivities showed a circadian rhythm with a trough at CT4 and a peak at CT16. Northern blot analysis of clock genes in the airway epithelium including the larynx also showed circadian rhythm (Fig. 4.17). Per1 and Per2 expressions were peaked in the daytime, while Bmal1 and Clock were peaked in the nighttime.

Oscillations were abolished in arrhythmic Cry1-/-Cry2-/- knockout mice and Clock mutant mice (Fig. 4.18). Lesioning of the master clock in the suprachiasmatic nucleus (SCN) in wild-type animals also abolished the rhythmic expression of Per1 and Per2 in the laryngeal and tracheal mucosa (Fig. 4.19). These findings indicate that respiratory system cells contain a functional peripheral oscillator that is controlled by the SCN.

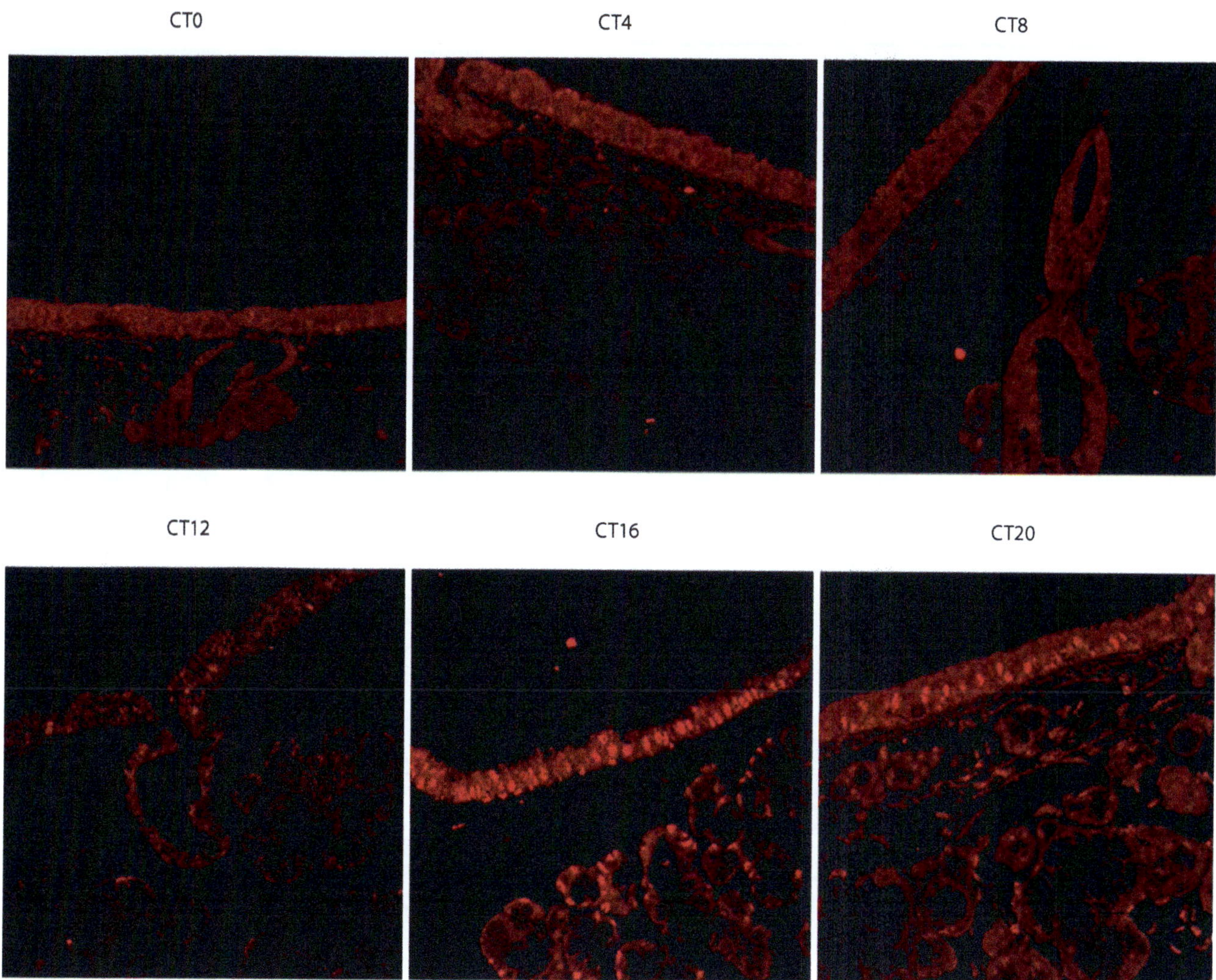

Fig. 4.16 Circadian expression of Per2 protein in the epithelium and submucosal glands in subglottic region was peaked at CT16. Per2 immunoreactivity was observed in the nucleus of epithelial cells and acinar cells in CT12–20, while no immunoreactivity was observed in CT0–8. Note the high level of nuclear immunohistochemical staining in tracheal epithelium and submucosal glands at CT16

In order to identify the route of signal transmission from SCN, we studied the effect of denervation on the expressions of Per1 and Per2 in the larynx. Unilateral vagotomy and bilateral sympathectomy were performed for wild-type mice. While sympathectomy did not affect the expression of Per1 and Per2, unilateral vagotomy significantly decreased the expression of Per1 and Per2 at CT16. At the laryngeal glands in vagotomized side, circadian expressions of Per1 and Per2 were completely abolished (Fig. 4.20). On the other hand, Per protein expression in the other side did not show any changes. These results indicate that signals from SCN are mainly transmitted by the vagal nerve. Thus, peripheral clock mediated circadian expression of muscarinic acethylcholine receptor proteins, and parasympathetic signaling between SCN and respiratory tissues are essential gears in conferring circadian "time" information to airway glands.

In the present study, we revealed that the clock gene expression in the airway is regulated by the central clock in the SCN via the vagal system, and at the same time the vagal tone is influenced by circadian clock through the transcriptional regulation of muscarinic receptor genes. The nocturnal worsening of the airway diseases could be solved by regulation of these molecules.

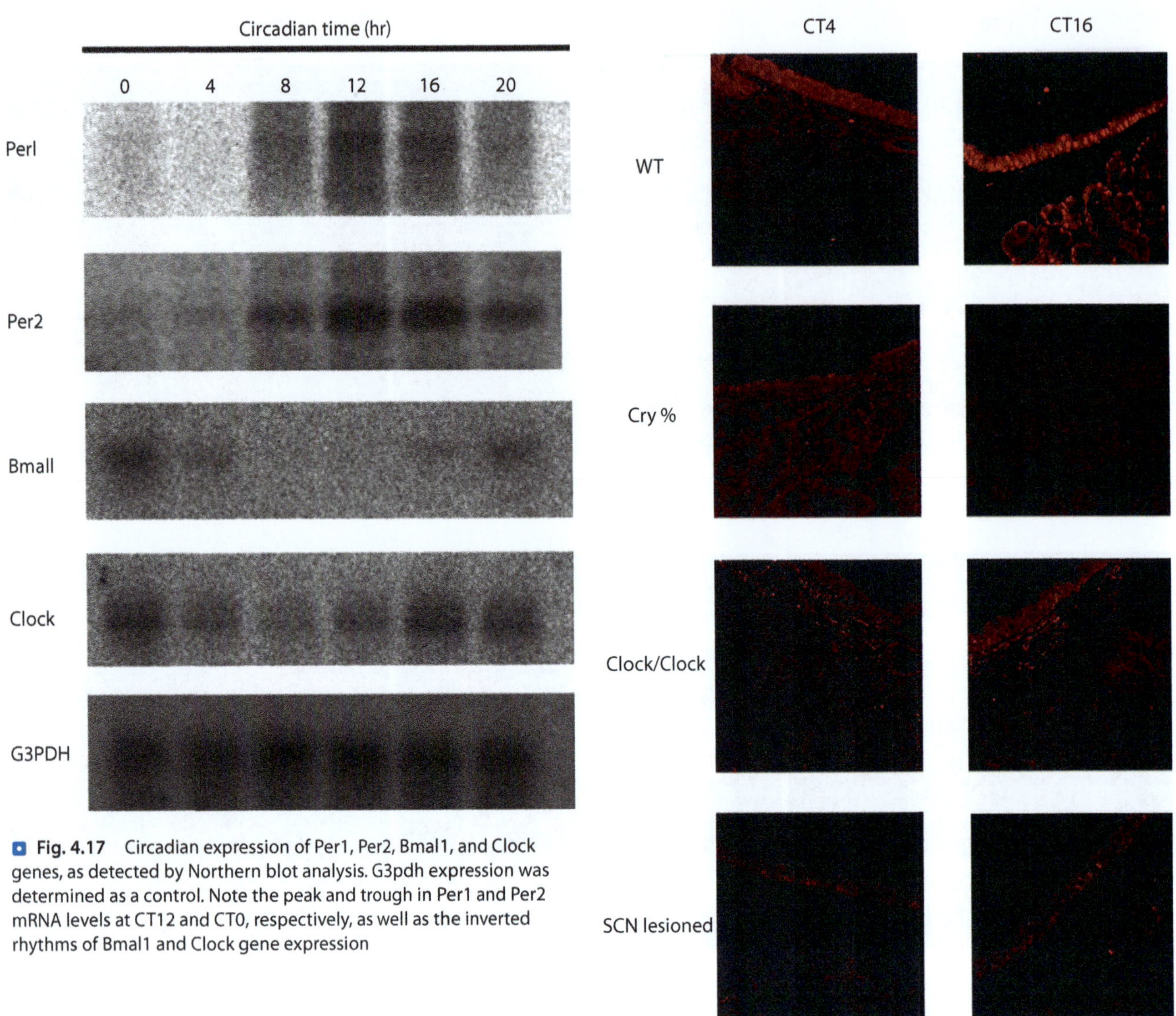

Fig. 4.17 Circadian expression of Per1, Per2, Bmal1, and Clock genes, as detected by Northern blot analysis. G3pdh expression was determined as a control. Note the peak and trough in Per1 and Per2 mRNA levels at CT12 and CT0, respectively, as well as the inverted rhythms of Bmal1 and Clock gene expression

Fig. 4.18 The circadian rhythm of Per2 protein expressions are abolished in the Cry-deficient mice, Clock mutant mice, and SCL-lesioned mice (*WT* wild type, *Cry-/-* Cry-deficient mice, *clock/clock* Clock mutant)

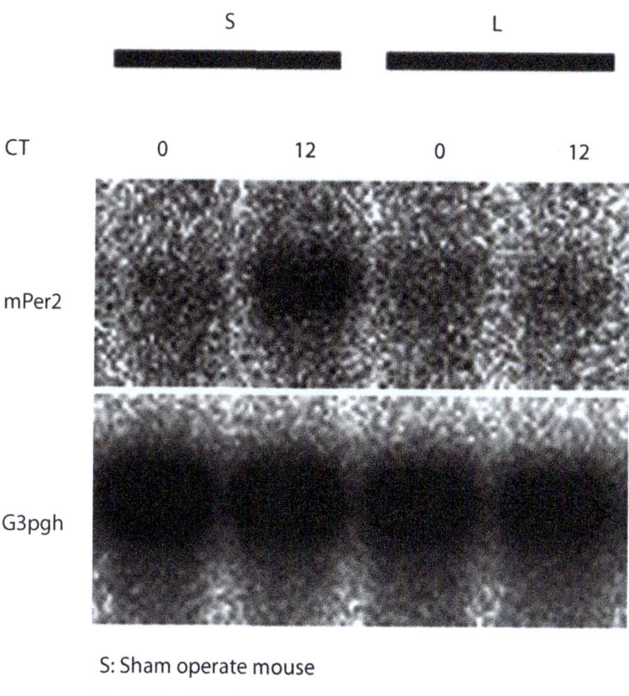

Fig. 4.19 Northern blot analysis showed that circadian rhythm of Per2 expression was abolished in SCN-lesioned mice. *S* sham-operated mice, *L* SCL-lesioned mice

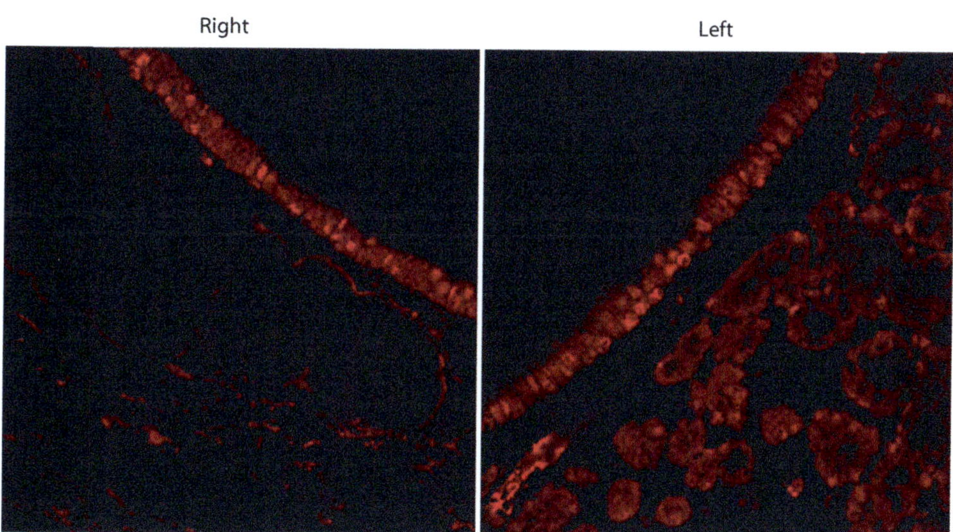

Fig. 4.20 While sympathectomized mice showed no alteration in mPER2 expression, in vagotomized mice mPER2 expression in submucosal gland on vagotomized side was significantly lower than the contralateral side and the sham-operated mice

References

1. Falck B, Hillarp NÅ, Thieme G, Torp A. Fluorescence of catecholamines and related compounds condensed with formaldehyde. J Histochem Cytochem. 1962;10:348–54.
2. Tranzer JP, Thoenen H. Electron microscopic localization of 5-hydroxydopamine (3,4,5-trihydroxyphenyl- ethylamine), a new 'false' sympathetic transmitter. Experientia. 1967;23:743–5.
3. Hisa Y. Fluorescence histochemical studies on the noradrenergic innervation of the canine larynx. Acta Anat. 1982;113:15–25.
4. Hisa Y, Koike S, Tadaki N, Bamba H, Shogaki K, Uno T. Neurotransmitters and neuromodulators involved in laryngeal innervation. Ann Otol Rhinol Laryngol Suppl. 1999;178:3–14.
5. Tanaka Y, Yoshida Y, Hirano M. Precise localization of VIP-, NPY- and TH-immunoreactivities of cat laryngeal glands. Brain Res Bull. 1995;36:219.
6. Hisa Y, Matui T, Fukui K, Ibata Y, Mizukoshi O. Ultrastructural and fluorescence histochemical studies on the sympathetic innervation of the canine laryngeal glands. Acta Otolaryngol. 1982;93:119–22.
7. Uno T, Hisa Y, Murakami Y, Okamura H, Ibata Y. Distribution of tyrosine hydroxylase immunoreactive nerve fibers in the canine larynx. Eur Arch Otorhinolaryngol. 1992;249:40–3.
8. Domeji S, Dahlqvist Å, Forsgren S. Enkephalin-like immunoreactivities in ganglionic cells in the larynx and superior cervical ganglion of the rat. Regul Pept. 1991;32:95–107.
9. Yoshida Y, Shimazaki T, Tanaka Y, Hirano M. Ganglions and ganglionic neurons in the cat's larynx. Acta Otolaryngol. 1993;113:415–20.
10. Shimazaki T. Morphological study of intralaryngeal ganglia and their neurons in the cat. Nippon Jibiinkoka Gakkai Kaiho. 1993;96:2044–56.
11. Tanaka Y, Yoshida Y, Hirano M. Ganglionic neurons in vagal and laryngeal nerves projecting to larynx, and their peptidergic features in the cat. Acta Otolaryngol Suppl. 1993;506:61–6.
12. Shimazaki T, Yoshida Y, Hirano M. Arrangement and number of intralaryngeal ganglia and ganglionic neurons: comparative study of five species of mammals. J Laryngol Otol. 1995;109:622–9.
13. Tooyama I, Kimura H. A protein encoded by an alternative splice variant of choline acetyltransferase mRNA is localized preferentially in peripheral nerve cells and fibers. J Chem Neuroanat. 2000;17:217–26.
14. Nakanishi Y, Tooyama I, Yasuhara O, Aimi Y, Kitajima K, Kimura H. Immunohistochemical localization of choline acetyltransferase of a peripheral type in the rat larynx. J Chem Neuroanat. 1999;17:21–32.
15. Nakajima K, Tooyama I, Yasuhara O, Aimi Y, Kimura H. Immunohistochemical demonstration of choline acetyltransferase of a peripheral type (pChAT) in the enteric nervous system of rats. J Chem Neuroanat. 2000;18:31–40.
16. Yasuhara O, Tooyama I, Aimi Y, Kimura H. Demonstration of cholinergic ganglion cells in rat retina: expression of alternative splice variant of choline acetyltransferase. J Neurosci. 2003;23:2872–81.
17. Masuko S, Kawasoe M, Chiba T, Shin T. Target-specific projection of intrinsic ganglionic neurons with different chemical codes in the canine larynx. Neurosci Res. 1991;9:270–8.
18. Tsuda K, Shin T, Masuko T. Immunohistochemical study of intralaryngeal ganglia in the cat. Otolaryngol Head Neck Surg. 1992;106:42–6.
19. Armstrong DM, Brady R, Hersh LB, Hayes RC, Wiley RG. Expression of choline acetyltransferase and nerve growth factor receptor within hypoglossal motoneurons following nerve injury. J Comp Neurol. 1991;304:596–607.
20. Avendano C, Umbriaco D, Dykes RW, Descarries L. Decrease and long-term recovery of choline acetyltransferase immunoreactivity in adult cat somatosensory cortex after peripheral nerve transections. J Comp Neurol. 1995;354:321–32.
21. Matsuura J, Ajiki K, Ichikawa T, Misawa H. Changes of expression levels of choline acetyltransferase and vesicular acetylcholine transporter mRNAs after transection of the hypoglossal nerve in adult rats. Neurosci Lett. 1997;236:95–8.
22. Tatemoto K. Neuropeptide Y: complete amino acid sequence of the brain peptide. Proc Natl Acad Sci U S A. 1982;79:5485–9.
23. Zukowska-Grojec Z, Karwatowska-Prokopczuk E, Rose W, Rone J, Movafagh S, Ji H, Yeh Y, Chen WT, Kleinman HK, Grouzmann E, Grant DS. Neuropeptide Y: a novel angiogenic factor from the sympathetic nerves and endothelium. Circ Res. 1998;83:187–95.
24. Bodanszky M, Klausner YS, Lin CY, Mutt V, Said SI. Synthesis of the vasoactive intestinal peptide (VIP). J Am Chem Soc. 1974;96:4973–8.
25. Muff R, Born W, Fischer JA. Adrenomedullin and related peptides: receptors and accessory proteins. Peptides. 2001;22:1765–72.
26. Domeji S, Dahlqvist Å, Forsgren S. Studies on colocalization of neuropeptide Y, vasoactive intestinal polypeptide, catecholamine-synthesizing enzymes and acetylcholinesterase in the larynx. Cell Tissue Res. 1991;263:495–505.
27. Kawasoe M, Shin T, Masuko S. Distribution of neuropeptide-like immunoreactive nerve fibers in the canine larynx. Otolaryngol Head Neck Surg. 1990;103:957–62.
28. Hisa Y, Uno T, Tadaki N, Murakami Y, Okamura H, Ibata Y. Distribution of calcitonin gene-related peptide nerve fibers in the canine larynx. Eur Arch Otorhinolaryngol. 1992;249:52–5.
29. Hisa Y, Uno T, Tadaki N, Koike S, Banba H, Tanaka M, Okamura H, Ibata Y. Relationship of neuropeptide to nitrergic innervation of the canine laryngeal glands. Regul Pept. 1996;66:197–201.
30. Rogers DF. Pharmacological regulation of the neuronal control of airway mucus secretion. Curr Opin Pharmacol. 2002;2:249–55.
31. Haga T. Molecular properties of muscarinic acetylcholine receptors. Proc Jpn Acad Ser B Phys Biol Sci. 2013;89:226–56.
32. Coulson FR, Fryer AD. Muscarinic acetylcholine receptors and airway diseases. Pharmacol Ther. 2003;98:59–69.
33. Rogers DF. Motor control of airway goblet cells and glands. Respir Physiol. 2001;125:129–44.
34. Vernino S, Hopkins S, Wang Z. Autonomic ganglia, acetylcholine receptor antibodies, and autoimmune ganglionopathy. Auton Neurosci. 2009;146:3–7.
35. Banes PJ. Circadian variation in airway function. Am J Med. 1985;79:5–9.
36. Bando H, Nishio T, van der Horst GTJ, Masubuchi S, Hisa Y, Okamura H. Vagal regulation of respiratory clocks in mice. J Neurosci. 2007;27:4359–65.
37. Nishio T, Bando H, Bamba H, Hisa Y, Okamura H. Circadian gene expression in murine larynx. Auris Nasus Larynx. 2008;35:539–44.
38. Tei H, Okamura H, Yamamoto S, et al. Circadian oscillation of a mammalian homologue of the Drosophila period gene. Nature. 1997;389:512–6.
39. Bae K, Jin X, Maywood ES, et al. Differential functions of mPer1, mPer2, and mPer3 in the SCN circadian clock. Neuron. 2001;30:525–36.
40. Zheng B, Albrecht U, Kaasik K, et al. Nonredundant roles of the mPer1 and mPer2 genes in the mammalian circadian clock. Cell. 2001;105:684–94.

Anatomy of Nerves

Recurrent Laryngeal Nerve

Toshiyuki Uno and Yasuo Hisa

T. Uno
Uno ENT Clinic, Tsuruga, Fukui 914-0052, Japan
e-mail: uno-iin@rm.rcn.ne.jp

Y. Hisa (✉)
Department of Otolaryngology-Head and Neck Surgery, Kyoto Prefectural University of Medicine, Kawaramachi-Hirokoji, Kamigyo-ku, Kyoto 602-8566, Japan
e-mail: yhisa@koto.kpu-m.ac.jp

Y. Hisa (ed.), *Neuroanatomy and Neurophysiology of the Larynx*, DOI 10.1007/978-4-431-55750-0_5

5.1 Introduction

It has long been known that the recurrent laryngeal nerve controls vocal cord motion. This was first discovered by Galenos (Galen in English) in the era of the Roman Empire. Galenos was born in Pergamum (currently in Turkey) in 129 AD and later became monarch Aurelius's doctor. In the era prior to Galenos, the brain was considered to harbor mental functions, while motions were controlled by the thoracic region. Galenos demonstrated that the brain controls motions, by showing in public that a struggling and grunting pig kept struggling without voice when the recurrent laryngeal nerve was cut [1]. The term "recurrens" was first used for the recurrent laryngeal nerve (nervus laryngeus recurrens) by the Belgian anatomist Vesalius in the sixteenth century; a description of the recurrent laryngeal nerve is found in his classic text entitled "Fabrica".

The recurrent laryngeal nerve has branches going to the esophagus and the trachea in the cervical region, eventually reaching the larynx. The terminal branch to the larynx is called the inferior laryngeal nerve. The inferior laryngeal nerve controls vocal cord motion by innervating four types of intrinsic laryngeal muscles other than the cricothyroid muscle (i.e., the thyroarytenoid, lateral cricoarytenoid, arytenoid, and posterior cricoarytenoid muscles). This nerve is known to contain autonomic nerve fibers [2] and sensory nerve fibers [3], in addition to motor nerve fibers.

The recurrent laryngeal nerve usually divides into two nerve branches (anterior and posterior), and the posterior branch forms Galen's anastomosis with the internal branch of the superior laryngeal nerve [4–8].

5.2 Nerve Fiber Composition

Many investigators have studied the fiber composition of the recurrent laryngeal nerve in order to indirectly confirm the issue of the conduction velocity of motor nerve fibers innervating intrinsic laryngeal muscles and the presence of mechanoreceptors such as muscle spindles in intrinsic laryngeal muscles. However, early studies failed to obtain consistent results because of species differences and various issues in the research methodology. However, studies carried out in the 1950s and thereafter revealed that the diameter of motor nerve fibers of the recurrent laryngeal nerve is generally smaller than that of the motor nerve fibers of limb muscles, although diameters vary markedly among motor nerve fibers in the recurrent laryngeal nerve (mostly 6–10 μm, with some thick fibers measuring 20 μm). This finding corroborates the observation that the conduction velocity (30–40 m/s) of the motor nerve fibers innervating the intrinsic laryngeal muscle is lower than that for limb muscles (50–60 m/s). The finding of only a few thick nerve fibers was also consistent with muscle spindles being scarce in intrinsic laryngeal muscles [9].

Gacek and Lyon [10], who used an electron microscope for their study, reported that the cat recurrent laryngeal nerve contains 565 and 482 myelinated nerve fibers on average on the right and left sides, respectively. They also reported that there were 827 and 680 unmyelinated nerve fibers on average on the right and left sides, respectively, although considerable variation among individual cats was observed. They speculated the reason why there were more myelinated nerve fibers on the right would be that sensory nerve fibers on the left terminate at the esophagus, while those on the right reach the trachea, terminating at the trachea and esophagus. They also carried out a nerve section experiment on the same occasion and speculated that unmyelinated nerve fibers in the recurrent laryngeal nerve would be sympathetic or parasympathetic fibers that have no relationships with motor function. The right-left difference in the number of nerve fibers was later examined in the rat by Dahlqvist et al. [11] and in humans and the giraffe by Harrison [8], and they reported that no such difference was detected. As to conduction velocity, the right-left difference in the thickness of nerve fibers has been studied. The left recurrent laryngeal nerve is longer than its right counterpart by 10 cm in human subjects [12], by 13 cm in the dog [13], and by 30 cm in the giraffe [8]. Therefore, based on the difference in conduction velocity, it is said that nerve fibers constituting the left recurrent laryngeal nerve are generally thicker than those on the right [12, 14].

5.3 Localizations of Nerve Fibers Innervating the Abductor and Adductor Muscles

The greatest interest in the field of laryngology from the end of the nineteenth century through the middle of the twentieth century focused on vocal cord position during recurrent laryngeal nerve paralysis. This interest was elicited by the report of Semon in 1881 [15]. He pointed out that nerve fibers innervating the abductor muscle are more subject to injury than nerve fibers innervating the adductor muscle in the case of recurrent laryngeal nerve injury and explained this by hypothesizing that fibers in the recurrent laryngeal nerve are arranged in a concentric fashion, with the nerve innervating the adductor muscle being located in the center [16]. However, in 1952, Sunderland and Swaney [17] morphologically studied the distributions of nerve fibers at various levels of the recurrent laryngeal nerve and reported that nerve fibers innervating the abductor muscle and those innervating the adductor muscle did not form separate fiber fascicles. The hypothesis of Semon regarding vocal cord position during recurrent laryngeal nerve paralysis was also later ruled out by various studies using the electrophysiological approach or other methods.

In recent years, Gacek et al. [18] have studied the distribution of labeled fibers in the recurrent laryngeal nerve electron microscopically at the level 1–2 cm caudal to the orifice of the larynx, after injecting horseradish peroxidase into the feline thyroarytenoid and posterior cricoarytenoid muscles. Their results made it apparent that nerve fibers innervating each of these muscles were scattered throughout the nerve fascicle. Therefore, nerve fibers innervating the abductor muscle and those innervating the adductor muscle were not located separately, instead being mixed, in the nerve fascicle.

5.4 Neurotransmitters Contained in the Recurrent Laryngeal Nerve

Malmgren and Gacek [19] classified cholinergic nerve fibers into two groups, in terms of the stainability and diameter of cat and human recurrent laryngeal nerve fibers as determined by acetylcholinesterase staining. One is a group of motor nerve fibers that were strongly stained and measured 4–12 μm, the other a group of nerve fibers measuring 1–5 μm in diameter with strong or moderate stainability, which the authors speculated were either sensory or autonomic nerve fibers. It is well known that motor nerve and parasympathetic preganglionic fibers are cholinergic nerve fibers. Our present study confirmed that recurrent laryngeal nerves are also cholinergic. The results of this study raise the possibility that cholinergic sensory nerve fibers are present as well.

In 1982, we first demonstrated, employing the Falck-Hillarp method, that adrenergic nerve fibers were contained in the canine recurrent laryngeal nerve [2]. In addition, in 1985, we also demonstrated for the first time, using an immunohistochemical method, that there were nerve fibers containing substance P (SP) [20]. Thereafter, Hauser-Kronberger et al. [21] reported that neuropeptide Y (NPY)-, vasoactive intestinal polypeptide (VIP)-, and calcitonin gene-related peptide (CGRP)-positive nerve fibers were contained in the human recurrent laryngeal nerve. Because it was difficult to identify neurotransmitters in the nerve fiber without employing ligation or crush processing, we later used gold-labeled cholera toxin B (CTBG) as a tracer and identified neurotransmitters contained in the innervating nerve fibers. As a result, it became apparent that the inferior laryngeal nerve, which is the terminal branch of the recurrent laryngeal nerve in the dog, has acetylcholine (Ach), noradrenaline (NA), CGRP-, SP-, NPY-, and nitric oxide (NO)-ergic nerve fibers [22].

5.5 Identification of NA Fibers in the Recurrent Laryngeal Nerve [2]

The canine inferior laryngeal nerve consists of the anterior and posterior branches at the laryngeal orifice, and the latter contains numerous unmyelinated fibers (Fig. 5.1). The canine inferior laryngeal nerve was crushed at the laryngeal orifice and processed by the Falck-Hillarp method. Although numerous NA nerve fibers were identified in the posterior branch (Fig. 5.2), only a few were found in the anterior branch. It was also noted that NA nerve fibers initially converged in the marginal region of the posterior branch and then branched off separately (Fig. 5.3a, b). These findings indicate that sympathetic nerve fibers are abundant in the posterior branch of the inferior laryngeal nerve. We can also reasonably speculate that these nerve fibers branch off from the main trunk of the nerve, while forming small nerve fascicles in the vicinity of the larynx, eventually reaching the muscle layer and the mucosa.

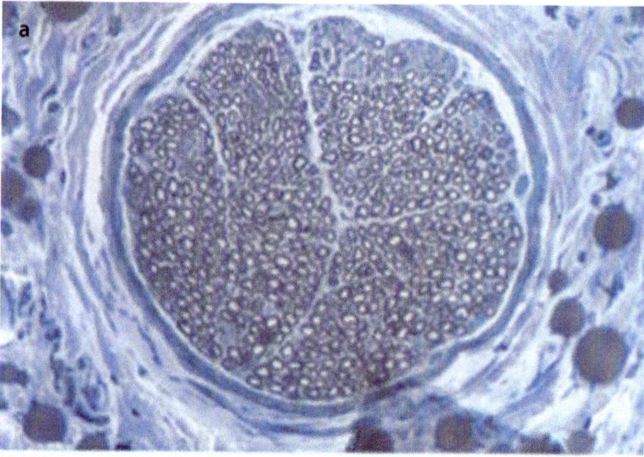

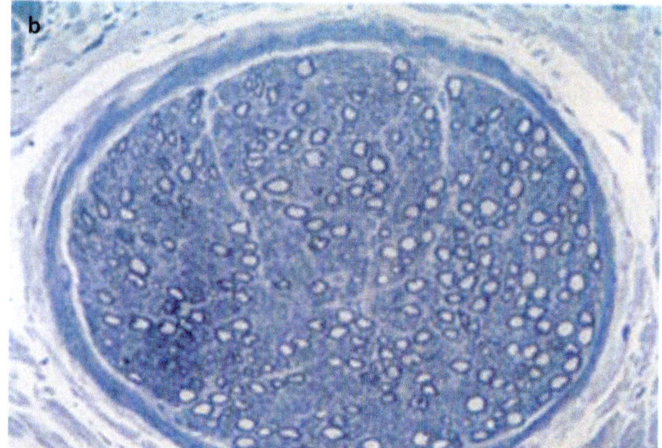

Fig. 5.1 Canine inferior laryngeal nerve stained with toluidine blue ((**a**) anterior branch, (**b**) posterior branch). The posterior branch can be seen to contain numerous unmyelinated fibers [2]

5.6 Immunohistochemical Identification of Neurotransmitters Contained in the Recurrent Laryngeal Nerve [22]

The canine inferior laryngeal nerve was crushed at the laryngeal orifice and subjected to immunohistochemical analysis using anti-SP antibody. No accumulation of SP-positive substances was found in the crushed area, but the presence of a few SP-positive nerve fibers was confirmed. In this regard, cell bodies extending fibers to the interior laryngeal nerve

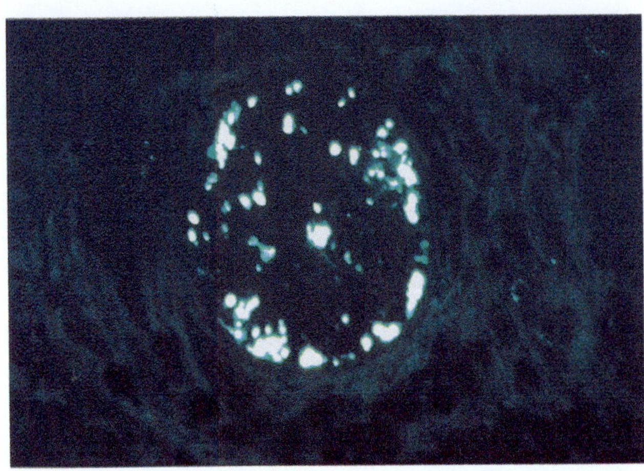

Fig. 5.2 Posterior branch of the canine inferior laryngeal nerve processed immunohistochemically by the Falck-Hillarp method. Several fluorescent fibers (NA nerve fibers) can be seen [2]

Table 5.1 Target substances and the results

NA	DMNV	NG	SCG
NADPHd	NADPHd	NADPHd[a]	NADPHd
ChAT[a]	ChAT[a]		
		TH[a]	TH[a]
CGRP[a]	CGRP[a]	CGRP[a]	CGRP
		SP[a]	SP
		ENK	ENK
		VIP	VIP[a]
		NPY	NPY[a]

NA nucleus ambiguus, *DMNV* dorsal motor nucleus of vagus nerve, *NG* nodose ganglion, *SCG* superior cervical ganglion
[a]Positive substance

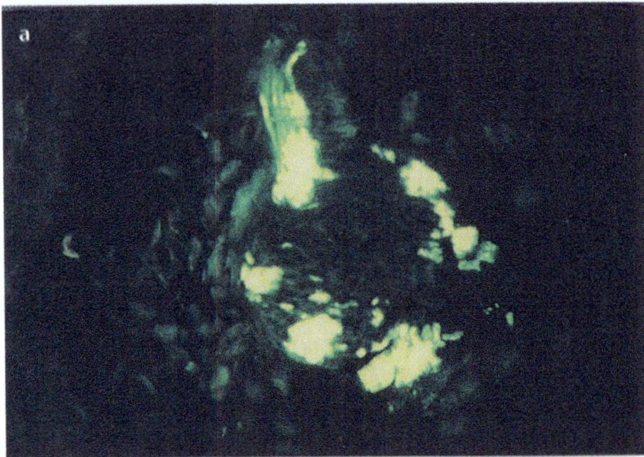

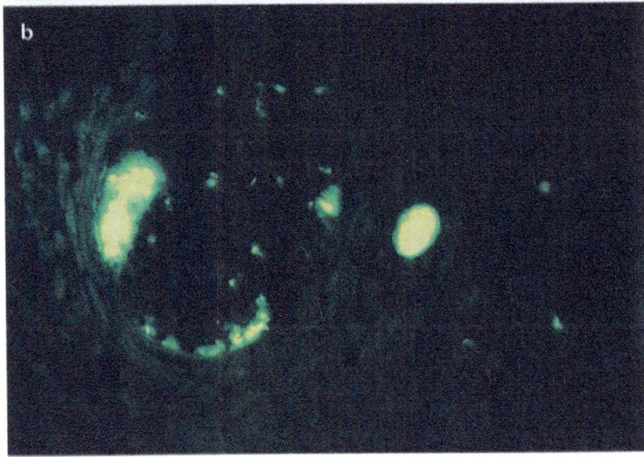

Fig. 5.3 Posterior branch of the canine inferior laryngeal nerve processed immunohistochemically by the Falck-Hillarp method. NA nerve fibers can be visualized as converging in the marginal region (**a**) and then branching off separately (**b**) [2]

were labeled using CTBG as a tracer, and labeled cells in the nucleus ambiguus, dorsal motor nucleus of the vagus nerve, nodose ganglion, and superior cervical ganglion were examined for neurotransmitters.

CTBG was injected into the canine right inferior laryngeal nerve at the level of the first tracheal ring and subjected to perfusion fixation after 48 hour. The brain stem, nodose ganglia, and superior cervical ganglia were extirpated, and sections were prepared. After visualizing CTBG by silver enhancement, immunohistochemical procedures with various antibodies and the NADPH-diaphorase (NADPHd) histochemical method for identification of NO were carried out. Anti-choline acetyltransferase (ChAT) antibody was used for identification of Ach, and anti-tyrosine hydroxylase (TH) antibody was used for identification of NA. Table 5.1 shows the target substances and the results of analysis at each site.

In the nucleus ambiguous, almost all CTBG-labeled cells were positive for ChAT, and most of the labeled cells were positive for CGRP.

The majority of cells in the dorsal motor nucleus of the vagus nerve were positive for ChAT. CTBG-labeled cells were mostly positive for ChAT, and there were also CGRP-positive cells.

CTBG-labeled cells were scattered throughout the nodose ganglion, and CGRP, SP, TH, and NADPHd-positive cells were present as well (Fig. 5.4). CGRP-positive cells were observed most frequently, followed by NADPHd-positive and SP-positive cells, whereas TH-positive cells were very rare.

In the superior cervical ganglion, CTBG-labeled cells were located mainly on the caudal and medial sides of the ganglion, and TH-positive cells were most frequent, followed by NPY-positive cells. Though VIP-positive cells were also present, they were very rare.

These results indicate that the inferior laryngeal nerve functions not only as a motor nerve but also has sensory and autonomic nerve functions mediated via various transmitters. This very interesting finding suggests that attention should not be focused solely on vocal cord motion disorders when considering the pathological condition of the recurrent laryngeal nerve paralysis.

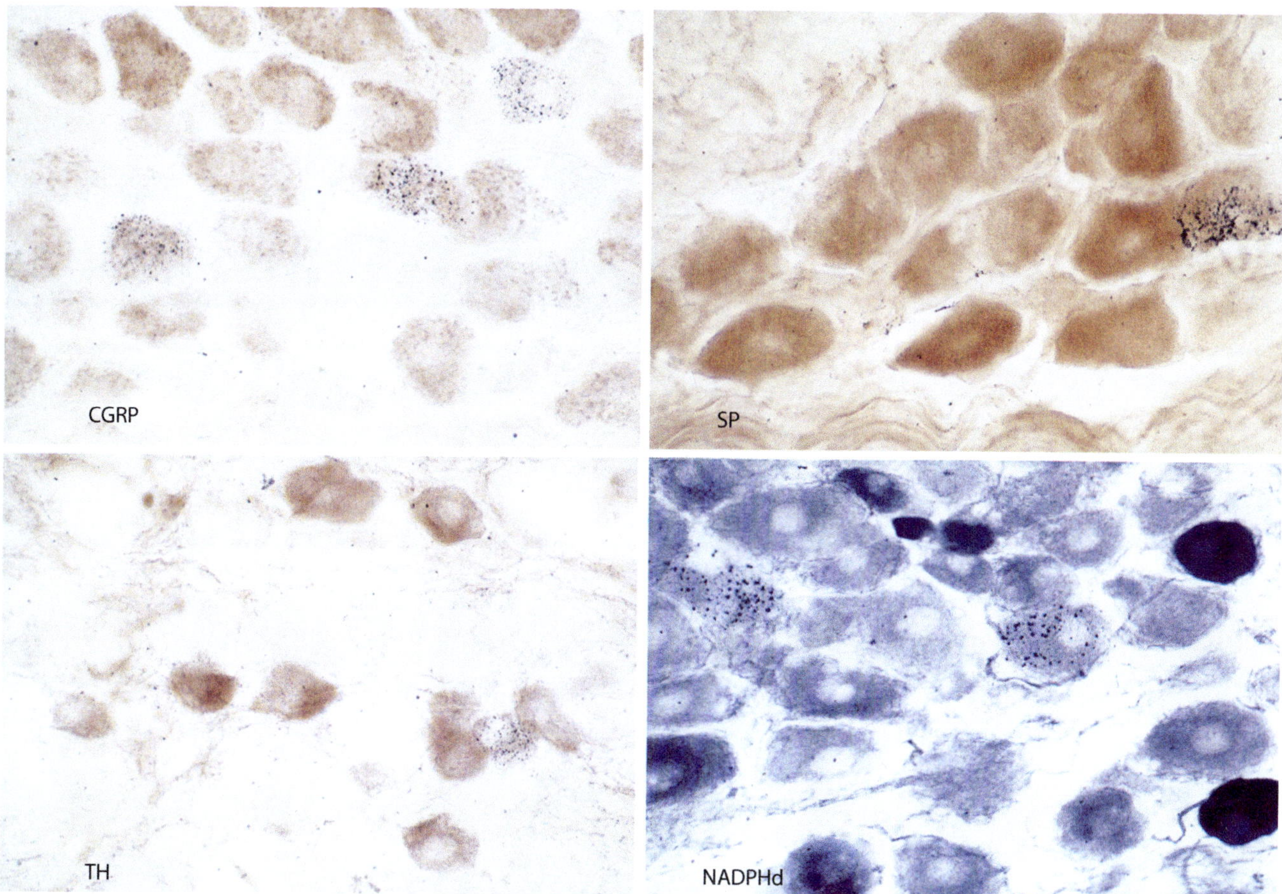

Fig. 5.4 Immunohistochemical methods using various antibodies were applied to labeled cells after injection of CTBG into the canine right inferior laryngeal nerve. CGRP, SP, TH, and NADPHd-positive cells can be seen

References

1. Freeman FR. Galen's ideas on neurological function. J Hist Neurosci. 1994;3:263–71.
2. Hisa Y. Fluorescence histochemical studies on the noradrenergic innervation of the canine larynx. Acta Anat. 1982;113:15–25.
3. Wyke BD, Kirchner JA. Neurology of the larynx. In: Hinchcliffe R, Harrison D, editors. Scientific foundations of otolaryngology. London: William Heinemann Medical Books; 1976. p. 546–74.
4. Dilworth TFM. The nerves of the human larynx. J Anat. 1921;56:48–52.
5. Lemere F. Innervation of the larynx. II. Ramus anastomoticus and ganglion cells of the superior laryngeal nerve. Anat Rec. 1932;54: 389–407.
6. Armstrong WG, Hinton JW. Multiple divisions of the recurrent laryngeal nerve. Arch Surg. 1951;62:532–9.
7. Bowden REM. The surgical anatomy of the recurrent laryngeal nerve. Br J Surg. 1955;43:153–63.
8. Harrison DFN. Fiber size frequency in the recurrent laryngeal nerves of man and giraffe. Acta Otolaryngol. 1981;91:383–9.
9. Boden RE. Innervation of intrinsic laryngeal muscles. In: Wyke B, editor. Ventilatory and phonatory control system. London: Oxford University Press; 1974. p. 370–91.
10. Gacek RR, Lyon MJ. Fiber components of the recurrent laryngeal nerve in the cat. Ann Otol Rhinol Laryngol. 1976;85:460–71.
11. Dahlqvist A, Carlsoo B, Hellstrom S. Fiber components of the recurrent laryngeal nerve of the rat: a study by light and electron microscopy. Anat Rec. 1982;204:365–70.
12. Shin T, Rabuzzi D. Conduction studies of the canine recurrent laryngeal nerve. Laryngoscope. 1971;81:586–96.
13. Atkins JP. An electromyographic study of recurrent laryngeal nerve conduction and its clinical application. Laryngoscope. 1973;83: 796–807.
14. Tomasch J, Britton WA. A fiber-analysis of the recurrent laryngeal nerve supply in man. Acta Anat. 1955;23:386–98.
15. Semon F. Clinical remarks on the proclivity of the abductor fibers of the recurrent laryngeal nerves to become affected sooner than the adductor fibers, or even exclusively, in cases of undoubted central or peripheral injury or disease of the roots or trunks of the pneumogastric, spinal accessory, or recurrent nerves. Arch Laryngol. 1881;2: 197–222.
16. Hirose H. Sir Felix Semon's original article about the recurrent laryngeal nerve paralysis. Otolaryngol Head Surg (Tokyo). 1967;39:1053–7.
17. Sunderland S, Swaney WE. The intraneural topography of the recurrent laryngeal nerve in man. Anat Rec. 1952;114:411–26.
18. Gacek RR, Malmgren LT, Lyon MJ. Localization of adductor and abductor motor nerve fibers to the larynx. Ann Otol Rhinol Laryngol. 1977;86:770–6.
19. Malmgren LT, Gacek RR. Acetylcholinesterase staining of fiber components in feline and human recurrent laryngeal nerve. Acta Otolaryngol. 1981;91:337–52.
20. Hisa Y, Sato F, Fukui K, Ibata Y, Mizukoshi O. Substance P nerve fibers in the canine larynx by PAP immunohistochemistry. Acta Otolaryngol. 1985;100:128–33.
21. Hauser-Kronberger C, Albegger K, Saria A, Hacker GW. Regulatory peptides in the human larynx and recurrent nerves. Acta Otolaryngol. 1993;113:409–13.
22. Hisa Y, Uno T, Tadaki N, Okamura H, Ibata Y. Neurotransmitters in the canine inferior laryngeal nerve. Larynx Jpn. 1994;6:117–21.

Superior Laryngeal Nerve

Toshiyuki Uno and Yasuo Hisa

T. Uno
Uno ENT Clinic, Tsuruga, Fukui 914-0052, Japan
e-mail: uno-iin@rm.rcn.ne.jp

Y. Hisa (✉)
Department of Otolaryngology-Head and Neck Surgery,
Kyoto Prefectural University of Medicine, Kawaramachi-Hirokoji,
Kamigyo-ku, Kyoto 602-8566, Japan
e-mail: yhisa@koto.kpu-m.ac.jp

Y. Hisa (ed.), *Neuroanatomy and Neurophysiology of the Larynx*, DOI 10.1007/978-4-431-55750-0_6

6.1 Introduction

Galenos was the first to describe three pairs of branches from the vagal nerve that innervate the larynx in his historical text entitled "De nervorum dissectione (Anatomy of the nerve)." These branches are considered to correspond to the inferior laryngeal nerve and the internal and external branches of the superior laryngeal nerve. It has been said that the internal branch of the superior laryngeal nerve mainly conveys sensation from the larynx and that the external branch of this nerve innervates the cricothyroid muscle. Our present study, however, provides evidence that the internal and external branches both contain sensory, motor, and autonomic nerve fibers.

The superior laryngeal nerve diverges from the vagal nerve just beneath the nodose ganglion and descends to ramify into the internal and external branches at the level of the middle pharyngeal constrictor muscle. The internal branch of the superior laryngeal nerve runs anteroinferiorly, pierces the thyrohyoid membrane together with the superior laryngeal artery, and ramifies into anterior and posterior fascicles at a level corresponding to the center of the epiglottis. The anterior fascicle mainly covers the glottis, piercing the lateral wall of the larynx through the gap between the internal and lateral muscles and distributing over the glottic mucosa. The posterior fascicle further bifurcates into a branch (middle branch) to the epiglottis, aryepiglottic fold, and arytenoids and another branch that runs toward the caudal side while ramifying into thin branches and distributing over the subglottic mucosa. The last branch of the posterior fascicle merges with the posterior fascicle of the inferior laryngeal nerve, forming Galen's anastomosis. The presence of an anastomotic branch between the middle branch and the arytenoid muscle branch of the inferior laryngeal nerve has also been reported [1]. The external branch of the superior laryngeal nerve descends along with the inferior pharyngeal constrictor muscle and innervates the cricothyroid muscle. An anastomotic branch is known to be present between the inferior laryngeal nerve and the external branch of the superior laryngeal nerve in humans [2–4]. Displacements for the site of anastomosis reportedly vary among laryngeal nerve branches, such as Galen's anastomosis [1].

6.2 Fiber Composition

It has been reported based on light microscopic studies that the number of myelinated fibers contained in the internal branch of the superior laryngeal nerve is 2,188–2,776 in the cat [5], about 15,000 in humans [6], and 317 and 354 on average on the right and left sides, respectively, in the rat, showing no right-left difference [7]. In recent years, advances in electron microscopic measurement and statistical analysis methods have enabled precise and accurate measurement of nerve fibers. Rosenberg et al. [8] reported the numbers of myelinated and unmyelinated fibers in the adult rat internal branch to be 335 ± 40 and 325 ± 62, respectively. Mortelliti et al. [9] reported the numbers of myelinated and unmyelinated fibers in the adult human internal branch to be $10{,}179 \pm 1{,}969$ and 10,469 ($n = 1$), respectively. In 1,955, Pressman and Kelemen [10] reviewed prior studies and stated that most of the internal branch consisted of myelinated fibers, with the proportion of unmyelinated fibers being very low. These findings indicate that the ratio of myelinated to unmyelinated fibers is essentially 1:1, at least in humans and dogs. Furthermore, the diameters of the myelinated and unmyelinated nerves in the rat internal branch were reported to be 2.92 ± 0.39 μm and 0.453 ± 0.035 μm, respectively [8].

There have been only a few reports on the fiber composition of the external branch of the superior laryngeal nerve. Domeij et al. [7] reported that the mean number of myelinated fibers in the rat external branch was 330 on the right side and 311 on the left side, showing no right-left difference, the same as in the internal branch. A comparison of rat internal and external branches showed that these two branches had nearly the same number of myelinated fibers, showing no bias in the distribution of myelinated and unmyelinated fibers. The diameters of myelinated and unmyelinated fibers were 0.5–12 μm and 0.1–2.3 μm, respectively. Thus, the internal and external branches are generally considered to be indistinguishable from each other in terms of the fiber composition alone [7].

The internal and external branches both have several small branches consisting of unmyelinated fibers in the vicinity of the larynx. In the superior laryngeal nerve in the laryngeal area, ganglion cells (paraganglia) are present in the form of clusters of up to 80 cells [11, 12]. (▶ See Chap. 7 on intralaryngeal ganglion.)

6.3 Function of the Superior Laryngeal Nerve

6.3.1 Sensory Nerve Fibers in the Superior Laryngeal Nerve

In regard to the sensory innervation of the internal branch, several reports have documented unilateral innervation covering the supraglottic area to the trachea [13] or unilateral innervation in almost all areas except for the subglottic posterior wall, which received bilateral innervation [14]. However, these were electrophysiological studies and failed to clearly identify areas of innervation due to their technical limitations. In 1986, Tanaka et al. [15] conducted a detailed investigation of the course and distribution of sensory nerve fibers in the cat larynx, using the horseradish peroxidase (HRP) method, and reported that the anterior fascicle of the superior laryngeal nerve internal branch was distributed to the laryngeal surface of the epiglottis and the aryepiglottic fold, while the middle branch was distributed to the aryepiglottic fold, arytenoid apex, posterior part, lateral part, laryngeal vestibule, and rostral surface of the vocal cord. They also reported that the posterior fascicle further divided into four branches which were distributed over the caudal surface of the right and left vocal cords, the mucosa in the subglottic space, the mucosa of the posterior crico-arytenoid muscle, and the hypopharyngeal mucosa, and that the internal branch of the superior laryngeal nerve innervated not only supraglottic but also subglottic areas, in the form of unilateral innervation in the supraglottic area and bilateral innervation with ipsilateral dominance in the subglottic area.

In 1968, Suzuki and Kirchner [16] reported that afferent fibers were contained in the cat external branch, controlling sensory innervation under the anterior commissure. Maranillo et al. [4] who conducted a detailed investigation of the course of the human external branch found that fibers distributed over the subglottic area and thyroarytenoid joint were present in approximately 50 % of individuals and speculated that they might be sensory fibers. Using the HRP method, we previously demonstrated sensory nerve fibers, whose cells of origin lay in the nodose ganglion, to be contained in the dog external branch [17].

6.3.2 Motor Nerve Fibers in the Superior Laryngeal Nerve

There has been ongoing controversy since the 1930s as to whether the internal branch contains motor nerves innervating the arytenoid muscle. Lemere [18] and Murtagh and Campbell [19] denied the presence of motor nerves in the internal branch in the dog based on anatomical findings, and Meurmann [20] and Williams [21] also reported that arytenoid muscle contraction in response to stimulation of the internal branch was not observed. In contrast, Vogel [22] identified motor nerve endings in the periphery of the internal branch in human subjects and advocated the theory of double innervation of the arytenoid muscle by the inferior and superior laryngeal nerves. Recently, Sanders and Mu [23] used Sihler's staining technique to study the internal branch of the human superior laryngeal nerve in detail and reported the presence of arytenoid muscle branches. We also previously found, using the nerve tracer technique, that motor nerve fibers were contained in the internal branch in the dog [24].

The external branch consists mainly of motor nerve fibers and controls the motion of the cricothyroid muscle. In the 1950s, Murtagh and Campbell [19] carried out an electrophysiological study and hypothesized that the cricothyroid muscle would be subject to double innervation by the external branch and the inferior laryngeal nerve, but their hypothesis is not currently supported. However, in humans, anastomotic branches between the external branch and the inferior laryngeal nerve (on one side or both sides) have been found at a frequency of 85 % [4]. Therefore, the possibility remains that a few motor nerve fibers are distributed from the inferior laryngeal nerve to the cricothyroid muscle.

6.3.3 Autonomic Nerve Fibers in the Superior Laryngeal Nerve

Sympathetic nerve fibers were formerly considered to be derived from the superior cervical ganglion and to run along the superior and inferior laryngeal artery and vein and then enter the larynx [25, 26], but accurate identification was lacking. In 1982, we reported for the first time that numerous noradrenaline (NA) fibers were contained in the inferior laryngeal nerve and the internal and external branches of the superior laryngeal nerve and that these nerve fibers were mainly distributed to the blood vessels in the larynx and laryngeal glands [27, 28].

Starting in the 1980s and extending into the 1990s, studies employing the nerve tracer method revealed that parasympathetic nerve fibers (preganglionic fibers) were contained in the superior laryngeal nerve in the rat [29], dog [30, 31], hamster [32], and guinea pig [33] and that their cell bodies were present in the dorsal motor nucleus of the vagus nerve. (See the section on the dorsal motor nucleus of the vagus nerve.)

6.4 Neuropeptides Contained in the Superior Laryngeal Nerve

Several neuropeptides have been shown to be involved in laryngeal sensory innervation [34–38]. We previously reported on the content of each neuropeptide in sensory nerve fibers contained in the internal branch of the superior laryngeal nerve [38].

6.5 Identification of Sympathetic Nerve Fibers in the Superior Laryngeal Nerve [27]

The superior laryngeal nerve contains numerous unmyelinated fibers. However, whether or not they are autonomic nerve fibers has yet to be clarified. In order to identify sympathetic nerve fibers contained in the superior laryngeal nerve, we crushed the internal and external branches of the canine superior laryngeal nerve at the site of laryngeal entry and then visualized NA fibers employing a fluorescent histochemical method (Falck-Hillarp method).

Three or four nerve fascicles were found in the internal branch of the superior laryngeal nerve, and all of these fascicles contained numerous discrete NA fibers (Fig. 6.1a). In addition, ramification of small nerve fascicles consisting of NA fibers alone was observed. NA fibers were found in the external branch of the superior laryngeal nerve, just as in the internal branch (Fig. 6.1b). From these findings, it was apparent that all branches of the superior laryngeal nerve have sympathetic nerve fibers, and it was inferred that these nerve fibers branched out from the main trunk in the vicinity of the larynx, distributing to the blood vessels and glands in the larynx.

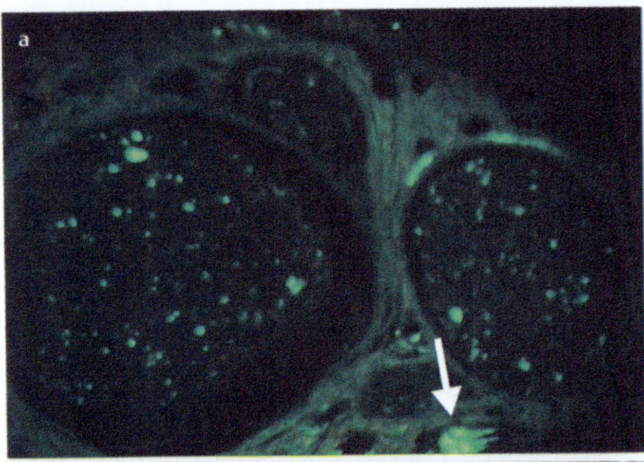

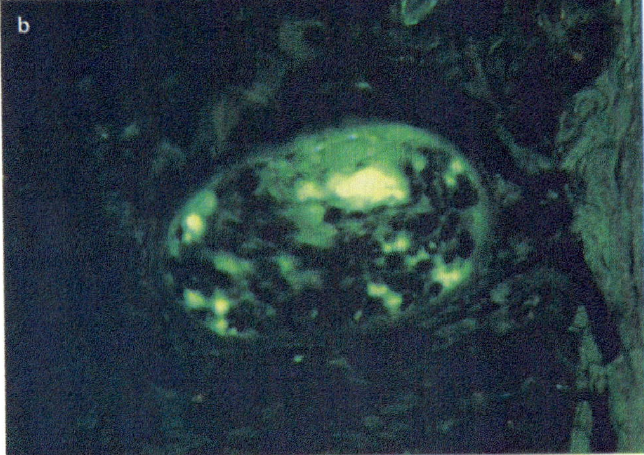

Fig. 6.1 Superior laryngeal nerve visualized by the Falck-Hillarp method (**a** internal branch, b external branch) [27]. (**a**) Nerve fascicles in the internal branch contain many sporadic NA fibers. There is a small nerve fascicle consisting of NA fibers only (→). (**b**) Numerous NA fibers are also contained in the external branch, just as in the internal branch

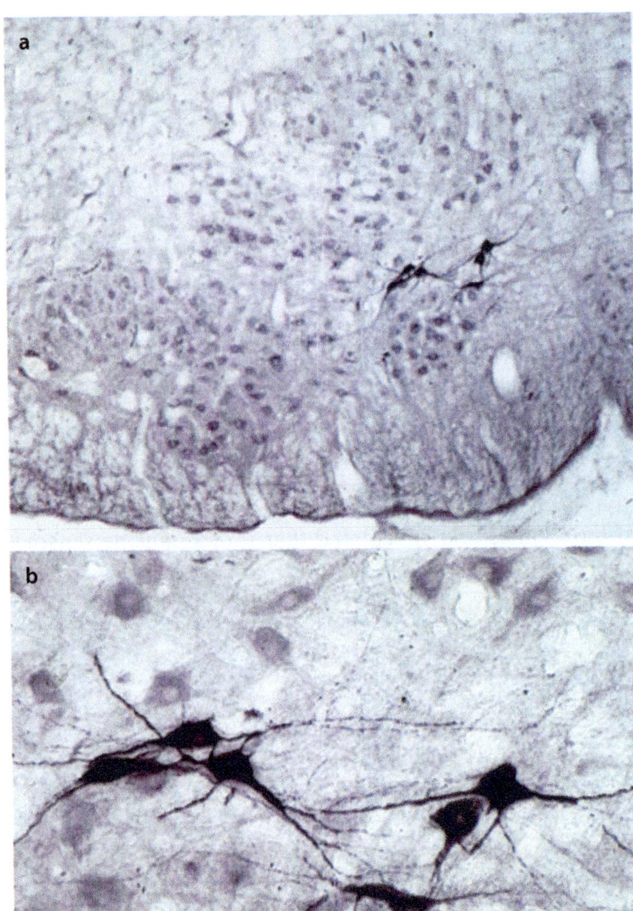

Fig. 6.2 The proximal end of the internal branch was soaked in CTB, and labeled cells were visualized by immunostaining [24]. (**a**) Labeled cells were present in the facial nerve nucleus (×70) (**b**) The labeled cells were multipolar and measured 35–50 μm in diameter (×350)

6.6 Localization of Cells of Origin of Motor Fibers Contained in the Internal Branch of the Superior Laryngeal Nerve [24]

The localization of cells of origin of nerve fibers contained in the internal branch was examined to determine whether motor nerves are contained in the internal branch of the superior laryngeal nerve. The internal branch of the canine superior laryngeal nerve on the right side was cut, and the proximal end was soaked in cholera toxin B (CTB) and subjected to perfusion fixation 2 days later. The brain stem was extirpated, and histological sections were prepared to visualize labeled cells by an immunohistochemical technique for CTB.

Although no labeled cells were found in the nucleus ambiguus, several labeled cells measuring 35–50 μm in diameter were observed in an area 4.1–7.4 mm rostral to the obex on the right ventrolateral side (Fig. 6.2a, b). This area corresponds to the region of the facial nucleus, indicating that motor nerve cells of the facial nucleus extend fibers to the internal branch of the superior laryngeal nerve. Further investigations are required to determine whether these cells control the intrinsic muscles of the larynx, such as the transverse arytenoid muscle, or the aryepiglottic muscle.

6.7 Involvement of Neuropeptides in Laryngeal Sensory Innervation [38, 39]

We investigated which neuropeptides play important roles in sensory innervation of the larynx. We labeled nodose ganglion cells that supply the internal branch of the superior laryngeal nerve with sensory nerve fibers, by infusing gold-labeled cholera toxin B (CTBG) into the internal branch of the canine superior laryngeal nerve. After colchicine processing followed by perfusion fixation, nodose ganglia were extirpated, and histological sections were prepared. After visualization of labeled cells by silver enhancement, immunohistochemical analyses were carried out using calcitonin gene-related peptide (CGRP), substance P (SP), and leu-enkephalin (ENK), and the positivity

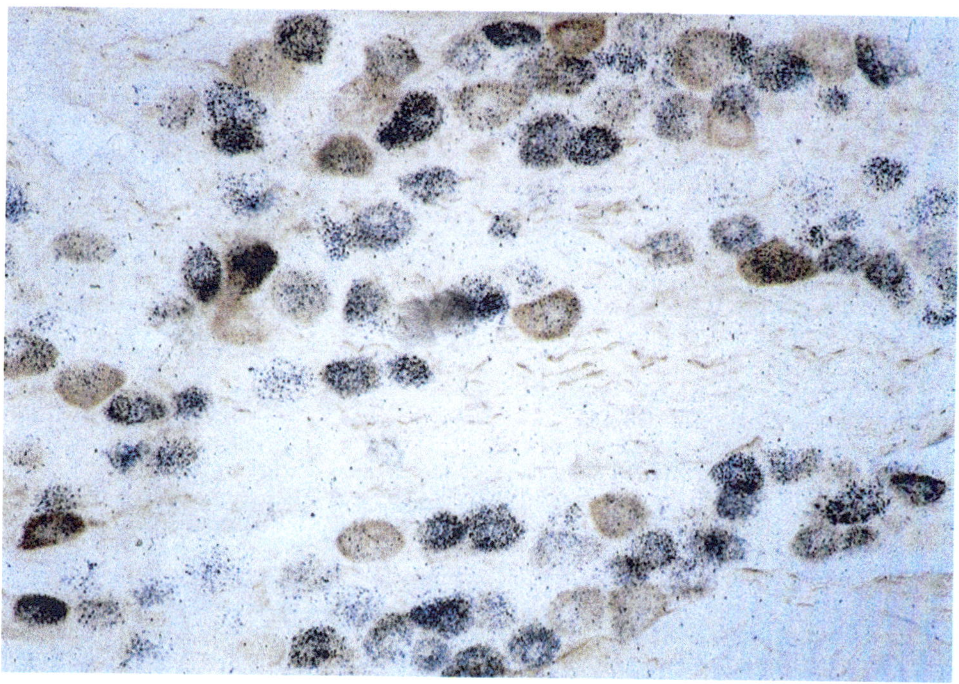

Fig. 6.3 CTBG was injected into the internal branch, and the SP immunohistochemical method was used to label cells in the nodose ganglion. There were several SP-positive cells among the CTBG-labeled cells [38]

rate of each neuropeptide in CTBG-labeled cells was determined.

Among all labeled cells, the CGRP positivity rate was the highest, 81.5 %, whereas the SP positivity rate was 24.5 %, and the ENK positivity rate was 7.0 % (Fig. 6.3).

CGRP and SP were the major neurotransmitters involved in laryngeal sensory innervation, and their distributions in the laryngeal mucosa were similar [34–38]. However, when CGRP and SP were compared in regard to the number of positive fibers, the results were inconsistent; we previously reported that the number of CGRP-positive fibers was higher than that of SP-positive cells [40], whereas others have found no difference between the two [35, 36]. Based on the results of our present study, nerve cells that extend fibers to the internal branch of the superior laryngeal nerve have an approximately threefold greater number of CGRP-positive cells than SP-positive cells. This finding supports the results of our study on the laryngeal mucosa and suggests that CGRP plays the most important role in sensory innervation of the larynx.

As to the neuropeptides contained in the nodose ganglion, cholecystokinin, neurokinin A, vasoactive intestinal polypeptide, and somatostatin, as well as CGRP and SP, have been reported [41, 42]. These observations suggest all of these peptides to play roles in laryngeal sensory innervation. Although this study also revealed the involvement of ENK in laryngeal sensory innervation, its low positivity rate, 7 %, suggests that its involvement may be restricted to the regulation and modification of neural transmission. This issue awaits further clarification.

References

1. Sanudo JR, Maranillo E, Leon X, Mirapeix RM, Orus C, Quer M. An anatomical study of anastomoses between the laryngeal nerves. Laryngoscope. 1999;109:983–7.
2. Wu BL, Sanders I, Mu L, Biller HF. The human communicating nerve. An extension of the external superior laryngeal nerve that innervates the vocal cord. Arch Otolaryngol Head Neck Surg. 1994;120:1321–8.
3. Kreyer R, Pomaroli A. Anastomosis between the external branch of the superior laryngeal nerve and the recurrent laryngeal nerve. Clin Anat. 2000;13:79–82.
4. Maranillo E, Leon X, Quer M, Orus C, Sanudo JR. Is the external laryngeal nerve an exclusively motor nerve? The cricothyroid connection branch. Laryngoscope. 2003;113:525–9.
5. Dubois F, Foley JO. Experimental studies on the vagus and spinal accessory nerves in the cat. Anat Rec. 1936;64:285–307.
6. Ogura JH, Lam RL. Anatomical and physiological correlations on stimulating the human superior laryngeal nerve. Laryngoscope. 1953;63:947–59.
7. Domeij S, Carlsoo B, Dahlqvist A, Hellstrom S, Kourtopoulos H. Motor and sensory fibers of the superior laryngeal nerve in the rat. A light and electron microscopic study. Acta Otolaryngol. 1989;108:469–77.
8. Rosenberg SI, Malmgren LT, Woo P. Age-related changes in the internal branch of the rat superior laryngeal nerve. Arch Otolaryngol Head Neck Surg. 1989;115:78–86.
9. Mortelliti AJ, Malmgren LT, Gacek RR. Ultrastructural changes with age in the human superior laryngeal nerve. Arch Otolaryngol Head Neck Surg. 1990;116:1062–9.
10. Pressman JJ, Kelemen G. Physiology of the larynx. Physiol Rev. 1955;35:506–54.
11. Domeij S, Carlsoo B, Dahlqvist A, Hellstrom S. Paraganglia of the superior laryngeal nerve of the rat. Acta Anat (Basel). 1987;130:219–23.
12. Yoshida Y, Shimazaki T, Tanaka Y, Hirano M. Ganglions and ganglionic neurons in the cat's larynx. Acta Otolaryngol. 1993;113:415–20.
13. Nagaishi T. Experimental studies about the vocal cord movement. Pract Otorhinolaryngol. 1938;33:518–725.
14. Suzuki M, Yamamoto W. Function of the recurrent laryngeal nerve. J Otolaryngol Jpn. 1971;74:454–5.
15. Tanaka Y. Distribution and pathways of peripheral sensory nerve fibers in the larynx and pharynx of cats. Otologia Fukuoka. 1986;32:1018–44.
16. Suzuki M, Kirchner JA. Afferent nerve fibers in the external branch of the superior laryngeal nerve in the cat. Ann Otol Rhinol Laryngol. 1968;77:1059–70.
17. Toyoda K. Localization of sensory neurons in the canine nodose ganglion sending fibers to the laryngeal nerves. J Otolaryngol Jpn. 1991;94:1888–97.

18. Lemere F. Innervation of the larynx. I. Innervation of the laryngeal muscles. Am J Anat. 1932;51:417–37.
19. Murtagh JA, Campbell CJ. Physiology of recurrent laryngeal nerve: report on progress. J Clin Endocrinol Metab. 1952;12:1398–401.
20. Meurmann OH. Theories of vocal cord paralysis. Acta Otolaryngol. 1950;38:460–72.
21. Williams AF. The recurrent laryngeal nerve and the thyroid gland. J Laryngol Rhinol Otol. 1954;68:719–25.
22. Vogel PH. The innervation of the larynx of man and the dog. Am J Anat. 1952;90:427–47.
23. Sanders I, Mu L. Anatomy of the human internal superior laryngeal nerve. Anat Rec. 1998;252:646–56.
24. Hisa Y, Uno T, Tadaki N, Murakami Y. Sensory, motor and autonomic nerve fibers of the internal branch of the canine superior laryngeal nerve. Trans Am Laryngol Assoc. 1992;113:98–103.
25. Ito C. Sympathetic-trophic innervation of the laryngeal muscles. J Otolaryngol Jpn. 1929;34:1207–28.
26. Sugano M. Experimental studies about the innervation of the laryngeal nerves. J Otolaryngol Jpn. 1930;35:1338–61.
27. Hisa Y. Fluorescence histochemical studies on the noradrenergic innervation of the canine larynx. Acta Anat (Basel). 1982;113:15–25.
28. Hisa Y, Matsui T, Fukui K, Ibata Y, Mizukoshi O. Ultrastructural and fluorescence histochemical studies on the sympathetic innervation of the canine laryngeal glands. Acta Otolaryngol. 1982;93:119–22.
29. Hinrichsen CF, Ryan AT. Localization of laryngeal motoneurons in the rat: morphologic evidence for dual innervation? Exp Neurol. 1981;74:341–55.
30. Wallach JH, Rybicki KJ, Kaufman MP. Anatomical localization of the cells of origin of efferent fibers in the superior laryngeal and recurrent laryngeal nerves of dogs. Brain Res. 1983;261:307–11.
31. Uno T. Autonomic neurons sending fibers into the canine laryngeal nerves -using a retrograde tracer technique with cholera toxin-. J Otolaryngol Jpn. 1993;96:66–76.
32. Hanamori T, Smith DV. Central projections of the hamster superior laryngeal nerve. Brain Res Bull. 1986;16:271–9.
33. Basterra J, Chumbley CC, Dilly PN. The superior laryngeal nerve: its projection to the dorsal motor nucleus of the vagus in the guinea pig. Laryngoscope. 1988;98:89–92.
34. Hisa Y, Sato F, Fukui K, Ibata Y, Mizukoshi O. Substance P nerve fibres in the canine larynx by PAP immunohistochemistry. Acta Otolaryngol. 1985;100:128–33.
35. Kawasoe M, Shin T, Masuko S. Distribution of neuropeptide-like immunoreactive nerve fibers in the canine larynx. Otolaryngol Head Neck Surg. 1990;103:957–62.
36. Tanaka Y, Yoshida Y, Hirano M, Morimoto M, Kanaseki T. Distribution of SP- and CGRP-immunoreactivity in the cat's larynx. J Laryngol Otol. 1993;107:522–6.
37. Hisa Y, Tadaki N, Uno T, Okamura H, Taguchi J, Ibata Y. Calcitonin gene-related peptide-like immunoreactive motoneurons innervating the canine inferior pharyngeal constrictor muscle. Acta Otolaryngol. 1994;114:560–4.
38. Hisa Y, Tadaki N, Uno T, Okamura H, Taguchi J, Ibata Y. Neuropeptide participation in canine laryngeal sensory innervation. Immunohistochemistry and retrograde labeling. Ann Otolaryngol Rhinol Laryngol. 1994;103:767–70.
39. Hisa Y, Koike S, Tadaki N, Bamba H, Shogaki K, Uno T. Neurotransmitters and neuromodulators involved in laryngeal innervation. Ann Otol Rhinol Laryngol Suppl. 1999;178:3–14.
40. Hisa Y, Uno T, Tadaki N, Murakami Y, Okamura H, Ibata Y. Distribution of calcitonin gene-related peptide nerve fibers in the canine larynx. Eur Arch Otorhinolaryngol. 1992;249:52–5.
41. Helke CJ, Hill KM. Immunohistochemical study of neuropeptides in vagal and glossopharyngeal afferent neurons in the rat. Neuroscience. 1988;26:539–51.
42. Helke CJ, Niederer AJ. Studies on the coexistence of substance P with other putative transmitters in the nodose and petrosal ganglia. Synapse. 1990;5:144–51.

Ganglion

Intralaryngeal Ganglion

Shinobu Koike and Yasuo Hisa

S. Koike
Department of Otolaryngology-Head and Neck Surgery, Kyoto Prefectural University of Medicine, Kawaramachi-Hirokoji, Kamigyo-ku, Kyoto 602-8566, Japan
e-mail: largo@koto.kpu-m.ac.jp

Y. Hisa (✉)
Department of Otolaryngology-Head and Neck Surgery, Kyoto Prefectural University of Medicine, Kawaramachi-Hirokoji, Kamigyo-ku, Kyoto 602-8566, Japan
e-mail: yhisa@koto.kpu-m.ac.jp

Y. Hisa (ed.), *Neuroanatomy and Neurophysiology of the Larynx*, DOI 10.1007/978-4-431-55750-0_7

7.1 Introduction

It has been generally accepted that parasympathetic control of the larynx originates in preganglionic neuronal bodies situated in the dorsal nucleus of the vagal nerve in the medulla oblongata, the axons of which reach the larynx through the superior or inferior laryngeal nerves and control laryngeal secretion and vascular tone. However, the location of the postganglionic neuronal bodies was a topic of some controversy. The existence of postganglionic parasympathetic neurons in the Auerbach plexus of the intestine has long been known, and thus it has been presumed that parasympathetic postganglionic neurons should exist close to the target organ. The intralaryngeal ganglia are regarded as most likely to be the parasympathetic ganglia because of their localization and for many other reasons.

7.2 Ganglia and Neuronal Cell Bodies in the Larynx

There have been multiple reports of ganglia existing in the larynx, and from characteristics of the ganglia reported, there seems to be three different groups of ganglia. Most of the reports have been on ganglia situated within or close to the nerve bundles of the superior laryngeal nerve. Since the first report in human and canine larynx by Elze [1] in 1923, many researchers have studied the distribution, size, and number of these ganglia using various methods [2–5]. These are the intralaryngeal ganglia that are the topic of this chapter.

Another type of ganglia is the "paraganglia" that Carlsoo [6] reported in the nerve bundles of the rat superior and inferior laryngeal nerves. The neuronal cell bodies in the "paraganglia" were reported to resemble type I or type II cells of the carotid body. None of these neurons were immunoreactive to vasoactive intestinal polypeptide (VIP) or encephalin (ENK) [7]. The neurons of these ganglia were also morphologically different from parasympathetic neurons known in the walls of the digestive tract and may be closer to the ganglia that Kummer and Neuhuber [8] reported later in the cardiac branch of the vagal nerve. However, ganglionic cells with VIP, ENK, or neuropeptide Y (NP-Y) have been reported in the vicinity of the "paraganglia" [7], so such cells may have immunoreactivity similar to that of the neurons in the intralaryngeal ganglia. The described location of the "paraganglia" is slightly rostral to the position of the intralaryngeal ganglia described above, but the existence of VIP- or ENK- or NP-Y-positive cells in the vicinity suggest that the "paraganglia" may be adjacent to or even be an island of neuroendocrine cells within the intralaryngeal ganglia.

A third type of neurons is found not in the mucosa or submucosa but between the muscle fibers of the intrinsic laryngeal muscles. These neurons are bipolar or pseudounipolar in shape [9] and differ from the neurons of the intralaryngeal ganglia in their localization and much smaller size of the ganglia they comprise. However, they include VIP-positive neurons [10] which is one characteristic they have in common with intralaryngeal ganglion neurons. (Details are available in Chap. 2 on intramuscular neurons in the intrinsic laryngeal muscles.)

7.3 Intralaryngeal Ganglion

Following the early reports mentioned above, Tsuda et al. [11] reported a detailed study on ganglia situated in the periventricular area along the internal branch of the superior laryngeal nerve in feline laryngeal mucosa. The neurons of the ganglia had ovoid cell bodies with a diameter of about 30 μm, and 90 % of the cells were positive to VIP immunohistochemistry.

In a study on cat larynx, Yoshida et al. [12] observed that a total of 600–800 nerve cells are included in the intralaryngeal ganglia, and based on results of immunohistochemistry on several neuropeptides and acetylcholine esterase (AchE) histochemistry, they suggested that they are cholinergic and thus parasympathetic in nature. Domeij et al. [13] showed that ganglionic cells in the rat larynx positive to AchE histochemistry, and so presumably parasympathetic, were also ENK-like immunoreactive.

Shimazaki [14] studied the intralaryngeal ganglia in the cat and reported that three to four large ganglia with 50–80 neurons each were found in the internal branches of the superior laryngeal nerve, while several small ganglia existed around the posterior cricoarytenoid muscle, and small ganglia with 15–25 neurons were seen close to the inferior laryngeal nerve. From the results of retrograde labeling experiments by injection of tracers into the area of the intralaryngeal ganglia, the existence of efferent innervation from sympathetic postganglionic neurons in the ipsilateral superior cervical ganglion and afferent innervation by sensory neurons in the ipsilateral nodose ganglion and the possibility of innervation from the neurons in the intralaryngeal ganglia to the superior cervical ganglion were suggested. The neurons in the intralaryngeal ganglia were positive to AchE histochemistry, while immunohistochemistry for various neurotransmitters showed that most of the neurons in the intralaryngeal ganglia were VIP positive and a small minority of them tyrosine hydroxylase (TH) or substance P (SP) positive, but none of them calcitonin gene-related peptide (CGRP) positive. Therefore, most of the neurons in the intralaryngeal ganglia are probably parasympathetic, but some of them have a sympathetic or sensory nature.

Because it is difficult to find material for the study of the normal human larynx, the intralaryngeal ganglia in humans have not been studied in detail. Existing studies were conducted on pathological larynges in some cases post chemotherapy with neurotoxic agents, and limitations on tissue preparations may have been reflected on the results of immunohistochemical analysis for neurotransmitters, making it difficult to interpret the results.

7.4 Neurotransmitters in Neurons of the Intralaryngeal Ganglion

We have studied the neurons of the intralaryngeal ganglia of the dog and rat using NADPH-diaphorase (NADPHd) histochemistry, which is a histochemical staining method for nitric oxide synthase that yields a very high contrast, and have also studied the localization of neuropeptides and gaseous neurotransmitters with immunohistochemical methods. The intralaryngeal ganglia contain both NADPHd-positive neurons and NADPHd-negative neurons, and a fine network of NADPHd-positive nerve fibers is seen surrounding the cell bodies of the NADPHd-negative cells (Fig. 7.1). The NADPHd-positive neurons are multipolar in shape [10, 15, 16]. Many of the neurons of the intralaryngeal ganglia are VIP positive [10] (Fig. 7.2), and many of the VIP-positive cells are also NADPHd positive [17]. Calcitonin gene-related peptide (CGRP), which is a typical neuropeptide known in sensory neurons, was not detected in any of the neurons by immunohistochemistry [10, 17]. Carbon monoxide (CO) is synthesized in cells by heme oxygenase as part of the heme metabolic pathway. Immunohistochemistry for heme oxygenase-2 (HO-2), which is a constituent isoform known to be localized in nervous tissue such as the brain, has shown that HO-2-positive neurons exist in the intralaryngeal ganglia [18]. Double staining techniques have shown that HO-2 and NADPHd are colocalized in some of the cells of the dog intralaryngeal ganglia.

Although NADPHd histochemistry is not isoform specific, the results of NADPHd histochemistry match the results of immunohistochemistry for neuronal nitric oxide synthase (nNOS) in the laryngeal nervous system [9, 15]. Therefore, nitric oxide (NO) is a potential gaseous neurotransmitter in the NADPHd-positive neurons in the intralaryngeal ganglia. Since many of the NADPHd-positive neurons in the intralaryngeal ganglia colocalized VIP but never CGRP, the CGRP-positive nerve fibers seen among the autonomic nerve fibers distributed around vessels and glands in the larynx must be extrinsic in origin and do not originate in the intralaryngeal ganglia. On the other hand, at least some of the nerve fibers with NO and VIP may originate in intralaryngeal ganglion neurons. Nonadrenergic noncholinergic (NANC) neurons are known as origins of nerve fibers controlling vascular tone, and VIP and NO are possible neurotransmitters involved, but since almost all the neurons in the intralaryngeal ganglia can be considered cholinergic as stated above, it is difficult to conceive that the intralaryngeal ganglia are the source of the NANC innervation.

The existence of HO-2-positive cells is another characteristic that is shared by intralaryngeal ganglia and the parasympathetic ganglia in the myenteric plexus of the digestive tract. HO-2 and nNOS were colocalized in some of the neurons of the canine intralaryngeal ganglia. The ratio of parasympathetic postganglionic neurons in the intestine that colocalize nNOS and HO-2 varies between species [19]. The ratio of neurons that colocalize HO-2 among the NADPHd-positive neurons in the canine myenteric plexus of the esophagus was 70–80 % in our study, and the ratio of colocalization of NADPHd reactivity among the HO-2-positive neurons in the same study was 35 % in the cervical and thoracic esophagus, but 53 % caudal to the diaphragm [20] (Fig. 7.3).

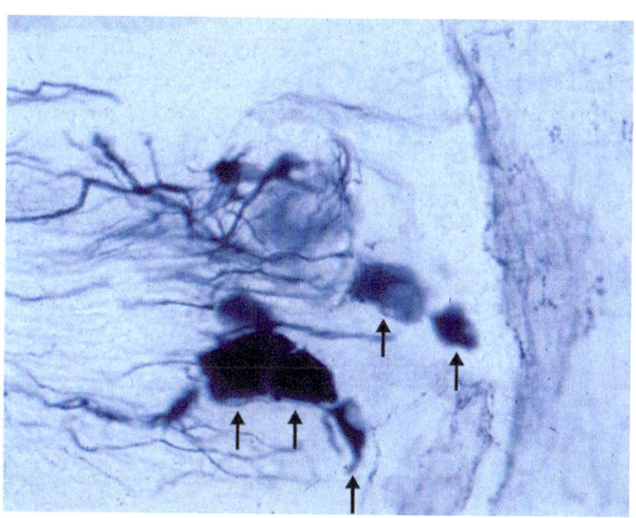

Fig. 7.1 NADPHd-positive cells (*arrows*) in rat intralaryngeal ganglion

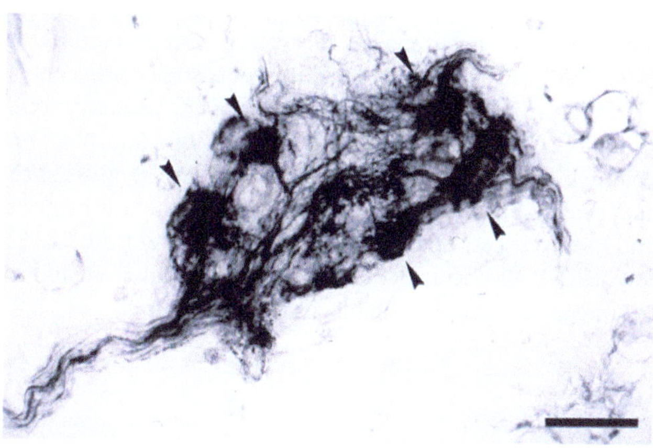

Fig. 7.2 Immunohistochemistry for VIP in canine intralaryngeal ganglion. This ganglion was found along a thick bundle of nerve fibers entering the cricothyroid muscle. Multipolar positive cells are observed (*arrow heads*) [10]

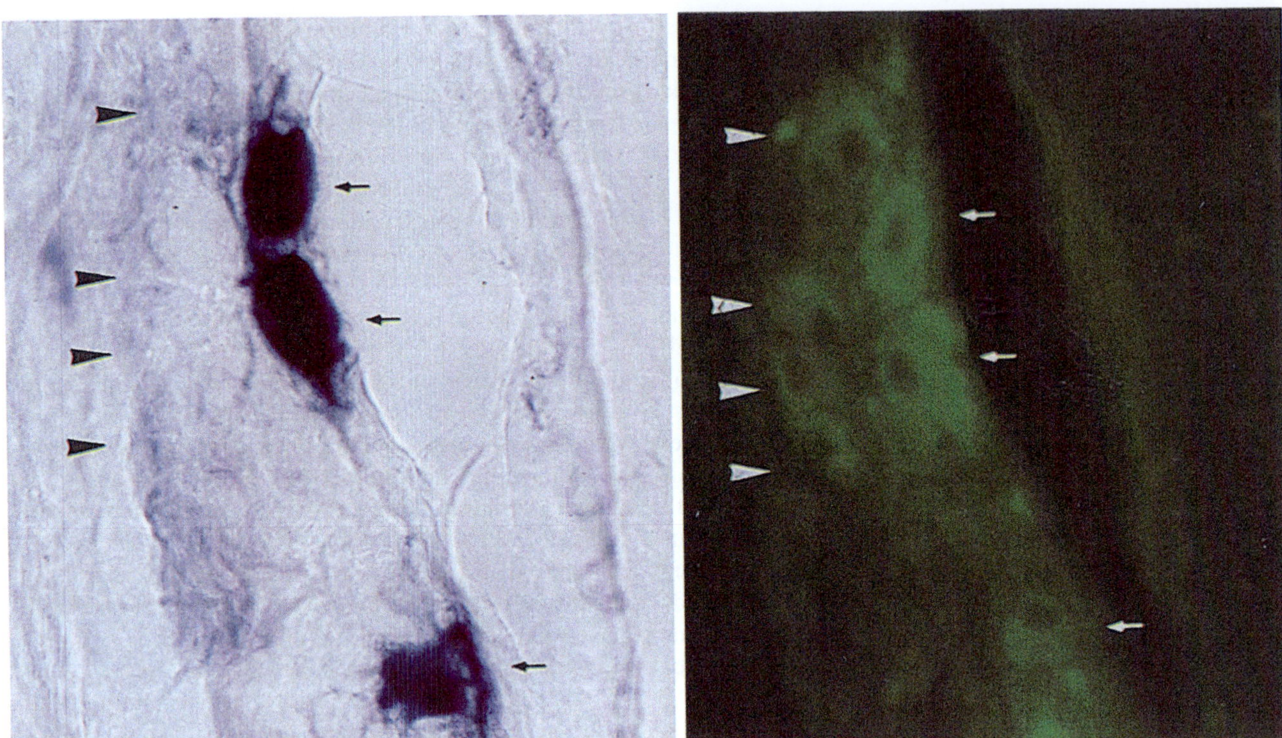

Fig. 7.3 Results of NADPHd histochemistry (*left*) and HO-2 immunohistochemistry (*right*) in nerve plexus in the canine cervical esophageal muscles. Cells which express both NADPHd and HO-2 are indicated by *arrows*. *Arrowheads* show the cells which express only HO-2 [20]

Colocalization of gaseous neurotransmitters is therefore another characteristic the neurons of the intralaryngeal ganglia and the neurons of the myenteric plexus of the esophagus have in common.

7.5 Capsaicin Receptors in the Neurons of the Intralaryngeal Ganglia

In a study of nociceptic receptors in the larynx, the authors have reported that transient receptor potential vanilloid 1 (TRPV1) and transient receptor potential vanilloid 2 (TRPV2) are also localized in neurons of the intralaryngeal ganglia [21]. These receptors are members of a family of ion channels that are activated not just by specific agonists such as capsaicin but also by various other stimuli such as protons and heat in excess of a specific threshold temperature (43 °C for TRPV1, 52 °C for TRPV2), thus with the potential ability to integrate various stimuli into their output [22, 23]. Mediators such as prostaglandins, adenosine, serotonin, bradykinin, and ATP that are released from surrounding damaged or inflamed tissues and from nociceptive neurons are known to cause hypersensitivities by lowering the temperature thresholds of the TRPV channels [24–27]. TRPV1 and TRPV2 were found to be colocalized in many of the neurons of the intralaryngeal ganglia [21] (Fig. 7.4) and may help integrate such various local inputs with parasympathetic preganglionic signals into their parasympathetic postganglionic output. In sensory ganglia, it is known that capsaicin receptors exist in the small dark neuronal somata that send thin unmyelinated C fibers to the periphery and large light A-type neurons which are the origin of thin myelinated Aδ fibers [28]. The existence of both TRPV1 and TRPV2 in the neurons of the intralaryngeal ganglia may be effective in processing various inputs that exist in the direct vicinity of the target tissue and providing necessary control of vascular tension or glandular secretion or triggering local inflammation.

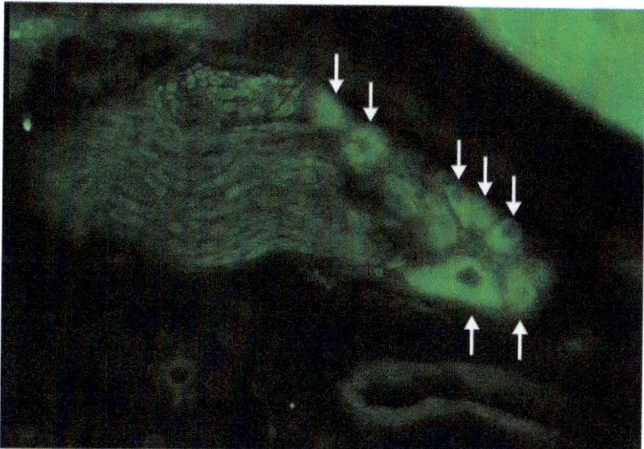

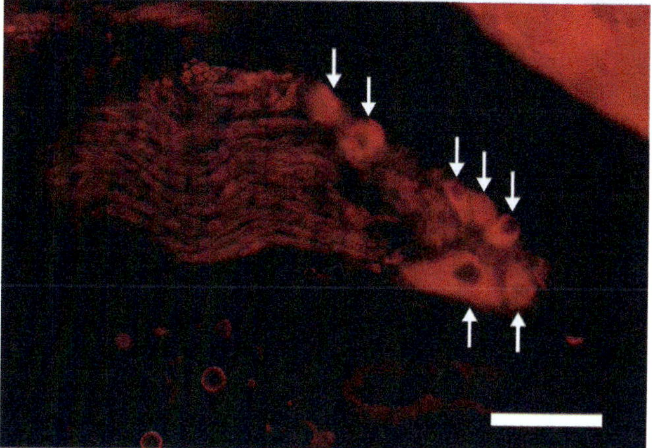

Fig. 7.4 Immunolocalization of TRPV1 (*upper*) and TRPV2 (*lower*) in the rat larynx. TRPV1 and TRPV2 are colocalized in the intralaryngeal ganglion [21]

References

1. Elze C. Kurze Mitteilung uber ein Ganglion in Nervus laryngeus superiorus des Menschen. Zeitschr f Anat u Entwicklungsgesch. 1923;69:630.
2. Nonidez JF. Innervation of the thyroid gland. II. Origin and course of the thyroid nerves in the dog. Am J Anat. 1931;48:299–329.
3. Lemere F. Innervation of the larynx. II. Ramus anastomoticus and ganglion cells of the superior laryngeal nerve. Anat Rec. 1932;54:389–402.
4. Afifi AB. The human epiglottis and its innervation. Arch Anat Histol Embryol. 1971;54:161–72.
5. Ramaswamy S, Kulasekaran D. The ganglion on the internal laryngeal nerve. Arch Otolaryngol. 1974;100:28–31.
6. Carlsoo B, Dahlqvist A, Domeij S, Hellstrom S, Dedo HH, Izdebski K. Carotid-body-like tissue within the recurrent laryngeal nerve: an endoneural chemosensitive micro-organ? Am J Otolaryngol. 1983;4:334–41.
7. Dahlqvist A, Forsgren S. Networks of peptide-containing nerve fibres in laryngeal nerve paraganglia: an immunohistochemical study. Acta Otolaryngol. 1989;107:289–95.
8. Kummer W, Neuhuber WL. Vagal paraganglia of the rat. J Electron Microsc Tech. 1989;12:343–55.
9. Hisa Y, Koike S, Uno T, Tadaki N, Tanaka M, Okamura H, Ibata Y. Nitrergic neurons in the canine intrinsic laryngeal muscle. Neurosci Lett. 1996;203:45–8.
10. Koike S, Hisa Y. Neurochemical substances in neurons of the canine intrinsic laryngeal muscles. Acta Otolaryngol. 1999;119:267–70.
11. Tsuda K, Shin T, Masuko S. Immunohistochemical study of intralaryngeal ganglia in the cat. Otolaryngol Head Neck Surg. 1992;106:42–6.
12. Yoshida Y, Shimazaki T, Tanaka Y, Hirano M. Ganglions and ganglionic neurons in the cat's larynx. Acta Otolaryngol. 1993;113:415–20.
13. Domeij S, Dahlqvist A, Forsgren S. Enkephalin-like immunoreactivity in ganglionic cells in the larynx and superior cervical ganglion of the rat. Regul Pept. 1991;32:95–107.
14. Shimazaki T. Morphological study of intralaryngeal ganglia and their neurons in the cat. Nippon Jibiinkoka Gakkai Kaiho. 1993;96:2044–56.
15. Hisa Y, Tadaki N, Uno T, Koike S, Tanaka M, Okamura H, Ibata Y. Nitrergic innervation of the rat larynx measured by nitric oxide synthase immunohistochemistry and NADPH-diaphorase histochemistry. Ann Otol Rhinol Laryngol. 1996;105:550–4.
16. Hisa Y, Koike S, Tadaki N, Bamba H, Shogaki K, Uno T. Neurotransmitters and neuromodulators involved in laryngeal innervation. Ann Otol Rhinol Laryngol Suppl. 1999;178:3–14.
17. Hisa Y, Uno T, Tadaki N, Koike S, Banba H, Tanaka M, Okamura H, Ibata Y. Relationship of neuropeptides to nitrergic innervation of the canine laryngeal glands. Regul Pept. 1996;66:197–201.
18. Koike S, Uno T, Bamba H, Shogaki K, Hirota R, Hisa Y. Localization of heme oxygenase-2 in the canine larynx. Acta Otolaryngol. 2001;121:315–7.
19. Miller SM, Reed D, Sarr MG, Farrugia G, Szurszewski JH. Haem oxygenase in enteric nervous system of human stomach and jejunum and co-localization with nitric oxide synthase. Neurogastroenterol Motil. 2001;13:121–31.
20. Bamba H, Uno T, Tamada Y, Tanaka M, Ibata Y, Hisa Y. Relationship between nitric oxide synthase and heme oxygenase-2 in the canine esophagus. Acta Histochem Cytochem. 2001;34:9–13.
21. Koike S, Uno T, Bamba H, Shibata T, Okano H, Hisa Y. Distribution of vanilloid receptors in the rat laryngeal innervation. Acta Otolaryngol. 2004;124:515–9.
22. Caterina MJ, Schumacher MA, Tominaga M, Rosen TA, Levine JD, Julius D. The capsaicin receptor: a heat – activated ion channel in the pain pathway. Nature. 1997;389:816–24.
23. Caterina MJ, Rosen TA, Tominaga M, Brake AJ, Julius D. A capsaicin-receptor homologue with a high threshold for noxious heat. Nature. 1999;398:436–41.
24. Julius D, Basbaum AI. Molecular mechanisms of nociception. Nature. 2001;413:203–10.
25. Mizumura K, Kumazawa T. Modification of nociceptic responses by inflammatory mediators and second messengers implicated in their action- a study in canine testicular polymodal receptors. Prog Brain Res. 1996;113:115–41.
26. Wood JN, Perl ER. Pain. Curr Opin Genet Dev. 1999;9:328–32.
27. Woolf CJ, Salter MW. Neuronal plasticity: increasing the gain in pain. Science. 2000;288:1765–9.
28. Holzer P. Capsaicin: cellular targets, mechanisms of action, and selectivity for thin sensory neurons. Pharmacol Rev. 1991;43:143–201.

Superior Cervical Ganglion

Hideki Bando, Shinji Fuse, Atsushi Saito, and Yasuo Hisa

H. Bando • S. Fuse • Y. Hisa (✉)
Department of Otolaryngology-Head and Neck Surgery,
Kyoto Prefectural University of Medicine, Kawaramachi-Hirokoji,
Kamigyo-ku, Kyoto 602-8566, Japan
e-mail: yhisa@koto.kpu-m.ac.jp

A. Saito
Department of Otolaryngology,
North Medical Center, Kyoto Prefectural University of Medicine,
Yosano-cho Otokoyama 481, Yosa-gun, Kyoto 629-2261, Japan

Y. Hisa (ed.), *Neuroanatomy and Neurophysiology of the Larynx*, DOI 10.1007/978-4-431-55750-0_8

8.1 Ganglion: Sympathetic Ganglion

Sympathetic innervation in the head and neck is delivered via the cervical sympathetic ganglions, and the preganglionic neurons are located in the spinal cord. The sympathetic regulations of the larynx including vasoconstriction of blood vessels and mucous secretion of laryngeal glands are also mediated by cervical sympathetic innervation. Little had been revealed about the laryngeal sympathetic innervation, until we started neuroanatomical study of the larynx in 1980s.

8.1.1 Cervical Sympathetic Ganglion

There are three pairs of cervical ganglion, the superior cervical ganglion (SCG), the medium cervical ganglion (MCG), and the stellatum ganglion (SG). The SCG is a linear-shaped ganglion and located ventral to transverse processes of the two topmost vertebrae, atlas and axis. The MCG is located at the level of the sixth vertebra. The SG is the complex of the inferior cervical ganglion and the first thoracic sympathetic ganglion and located dorsal to the subclavian artery at the level of the transverse process of seventh vertebra [1].

Neurons of sympathetic ganglion in mammals including human are generally multipolarized cell, which have numbers of dendrites with various sizes and a single axon with smooth outline and no branches. The total number of cells in the SCG is around a million in human and 16,000 in guinea pig [2]. The size of cell bodies is reported to be 20–50 μm in human and around 48 μm in cat [3]. We have reported that the cell size of canine SCG is about 30 μm [4].

Acetylcholine released from preganglionic nerves in the sympathetic ganglions activates nicotinic acetylcholine receptors on postsynaptic nerves. In response to this stimulus, postganglionic neurons release norepinephrine, which activates adrenergic receptors on the peripheral target organs.

8.1.2 Neuropeptides in the Sympathetic Ganglion Cell

Recent studies have revealed that sympathetic ganglion cells express not only the classical neurotransmitters such as noradrenaline and acetylcholine but various neuropeptides which play roles as neurotransmitter or neuromodulator. In the sympathetic ganglion, the expressions of peptides such as calcitonin gene peptide (CGRP) [5, 6], substance P (SP) [7], vasoactive intestinal peptide (VIP) [8], encephalin (ENK) [9, 10], and neuropeptide Y (NPY) [11] have been confirmed. These peptides are considered to be involved in the regulation of blood flow and gland secretion according to recent studies. These study reports that VIP, SP, and CGRP are involved in vasodilation and SP acts on vasodilation and vascular permeability. On the other hand, NPY is reported to affect vasoconstriction.

8.2 Sympathetic Projection to the Larynx

Although sympathetic innervation of the larynx was studied in the previous studies, the postganglionic projection to the larynx has not been clarified. Conventionally, postganglionic sympathetic nerves from SCG had been considered to run along superior and inferior laryngeal artery and vein. Our study identified the NA-positive fibers in the internal and external branch of superior laryngeal nerve and the inferior laryngeal nerve, which project to laryngeal vessels and glands [12, 13]. We have also revealed the origin of the sympathetic fibers of laryngeal nerves (superior laryngeal nerve and inferior laryngeal nerve), and most of the fibers were originated from SCG [4]. The distribution and the number of these neurons in SCG were also clarified. Moreover, we identified the co-expression of CGRP and neutral nitric oxide synthase (nNOS) in order to illuminate the projection of non-adrenergic, non-cholinergic nerve (NANC) fibers to the larynx [14].

8.2.1 Origin of the Sympathetic Fibers in the Laryngeal Nerves

The origin of the sympathetic fibers in the canine laryngeal nerves in the cervical sympathetic ganglions was analyzed [4]. Each laryngeal nerve was cut, and the proximal end was soaked in the solution of cholera toxin B (CTB): (a) internal branch of superior laryngeal nerve, (b) external branch of superior laryngeal nerve, and (c) inferior laryngeal nerve.

Four days later, immunohistochemistry for CTB was performed following perfusion and fixation.

Results

1. Medium cervical ganglion (MCG): No immunoreactivity was confirmed.
2. Stellatum ganglion (SG): No immunoreactivity was confirmed.
3. Superior cervical ganglion (SCG)
   (a) Internal branch of superior laryngeal nerve: There was a large number of CTB immunoreactive cells compared to the other two groups (◘ Fig. 8.1a). These were multipolarized cell with a single axon originated in the cell body and located in the internal and caudal region of the ganglion (◘ Fig. 8.1b).
   (b) External branch of superior laryngeal nerve: A small number of immunoreactive cells was confirmed in the internal and caudal region (◘ Fig. 8.2a).
   (c) Inferior laryngeal nerve: Immunoreactive cells were confirmed also in the medial and caudal region (◘ Fig. 8.2b).

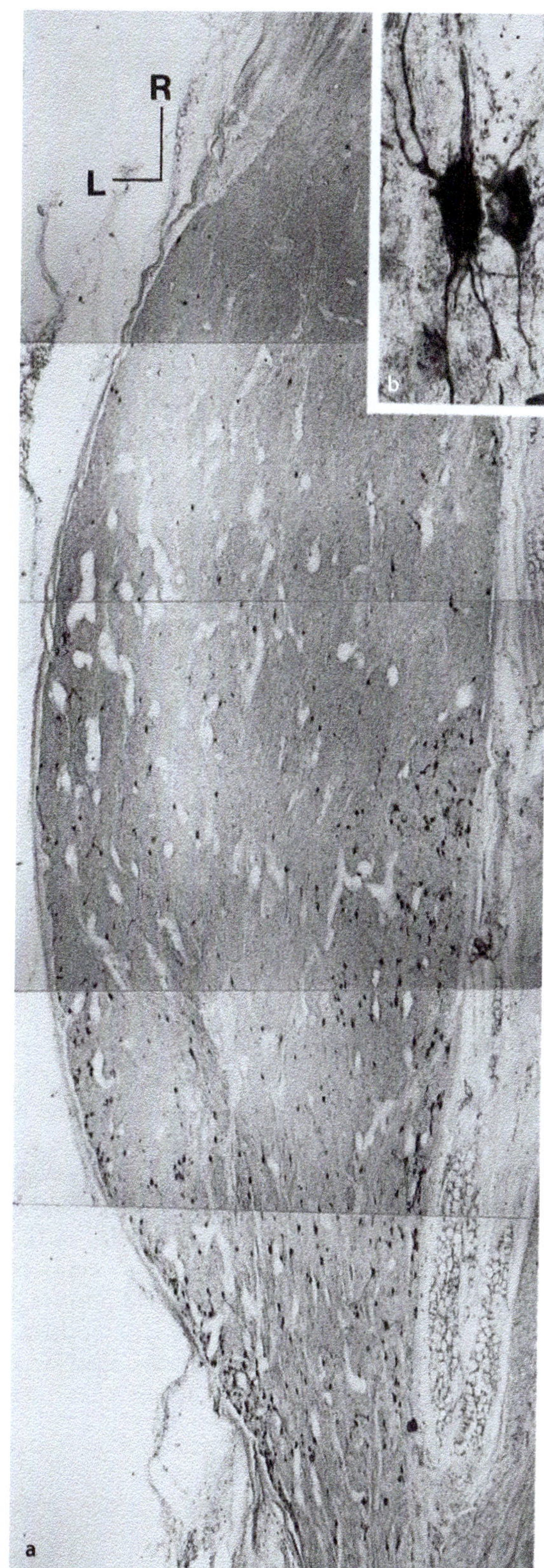

Fig. 8.1 Canine SCG (*right*) cells conjugated with cholera toxin B injected into the internal branch of superior laryngeal nerve (*L* lateral, *R* rostal). (**a**) Cholera toxin B-conjugated cells are distributed mainly in the caudal and median part of the ganglion. (**b**) Most of the conjugated cells are multipolar neuron with multiple ramified dendrites and a single axon

This study revealed that the cell body of sympathetic fibers in the laryngeal nerve (the internal and external branch of superior laryngeal nerve and the inferior laryngeal nerve) is located mainly in the SCG, especially in the medial and caudal region. The number of CTB-positive cells in the group of internal branch of superior laryngeal nerve denervation was as 20 times more than the other two groups, which indicate that most of the sympathetic nerve fibers are projected to the larynx via the superior laryngeal nerve.

Yoshida et al. reported that some of the laryngeal sympathetic nerve fibers in the cat were innervated via MCG. We have speculated that the difference of the results is caused by the immaturity of MCG in the young dogs used in our study.

8.2.2 CGRP Immunoreactive Cells in SCG

It is known that CGRP is distributed in parasympathetic ganglions and plays important roles in autonomic controls of the gland secretion and the peripheral blood flow. Although several studies have identified CGRP-positive cells in sympathetic ganglions [15–17], the roles in sympathetic nervous system have not been illuminated.

We conducted immunohistochemical study for CGRP in canine SCG, and CGRP-positive cells with a number of dendrites were distributed diffusely. The size of the cell was about 25 μm uniformly. As small intensely fluorescent (SIF) cells are about 6–12 μm in diameter, CGRP-positive cell was considered to be the principal ganglion cell. CGRP-positive cells are diffusely distributed all over the ganglion (Fig. 8.3). The number of the CGRP-positive cell was about 8,000 in a ganglion, which amounts to about 8 % of whole ganglion cells. CGRP-positive cells were confirmed in cat [15], dog [16], and human [17], while no immunoreactivity was detected in rat [18] and guinea pig [16]. CGRP is concerned in the blood flow regulation as a strong vasodilation factor in the autonomic nervous system [19–21], and it is well known that CGRP is distributed in the parasympathetic ganglion [22]. Although the existence of CGRP immunoreactive neurons in the peripheral sympathetic nervous system is revealed, the significance of these results has not been unveiled.

8.2.3 Coexpression of CGRP and Nitric Oxide in SCG

While nitric oxide (NO) is known as a member of NANC neurotransmitters [23], the role in the laryngeal gland secretion has not been illuminated. Our previous study revealed that there are a number of nitrergic neurons around laryngeal glands, which play important roles in regulation of the gland secretion [24, 25]. We have also confirmed coexpression of VIP and CGRP in these fibers [25]. Although a number of NO-positive cells were identified in

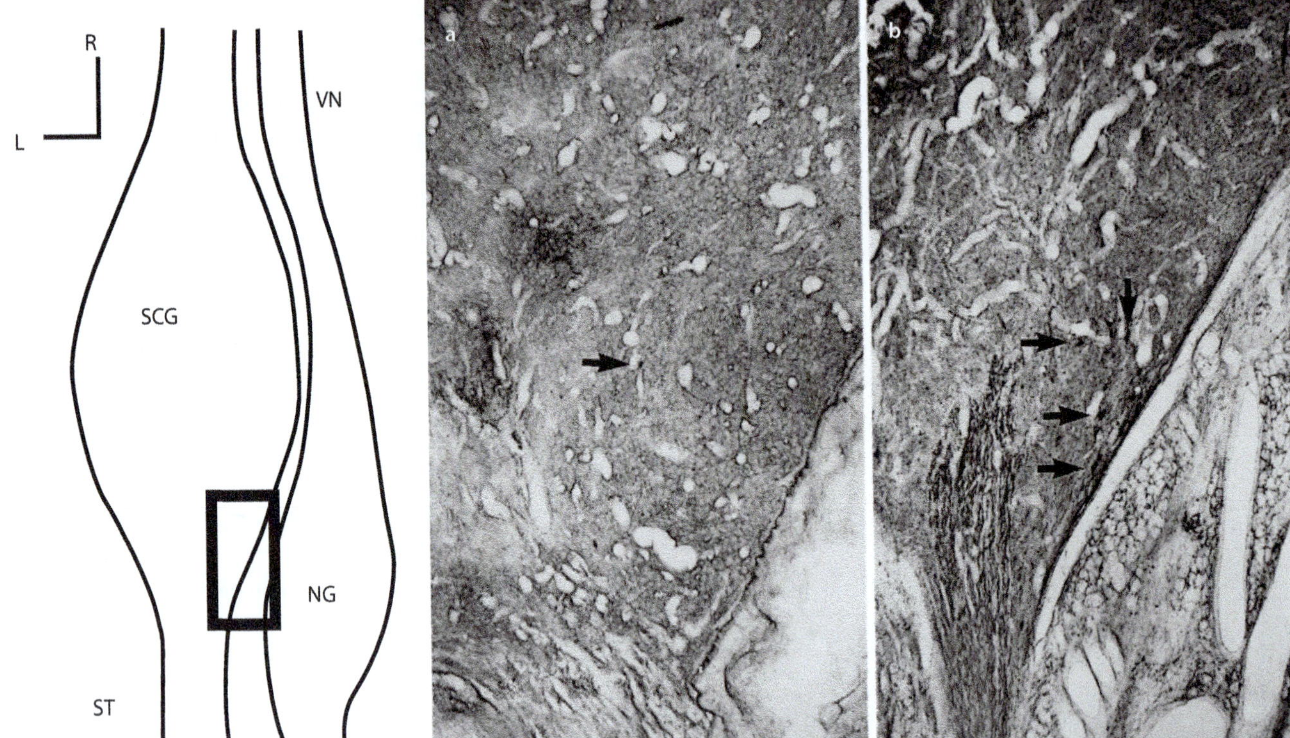

Fig. 8.2 Right canine SCG with conjugated CTB injected in external branch of superior laryngeal nerve (**a**) and inferior laryngeal nerve (**b**). (*L* lateral, *R* rostal, *NG* nodose ganglion, *SCG* superior cervical ganglion, *ST* sympathetic trunc, *VN* vagal nerve). (**a**) A single conjugated cell is found in caudal median region of SCG (*arrow*). (**b**) Small numbers of conjugated cells are observed in caudal median region

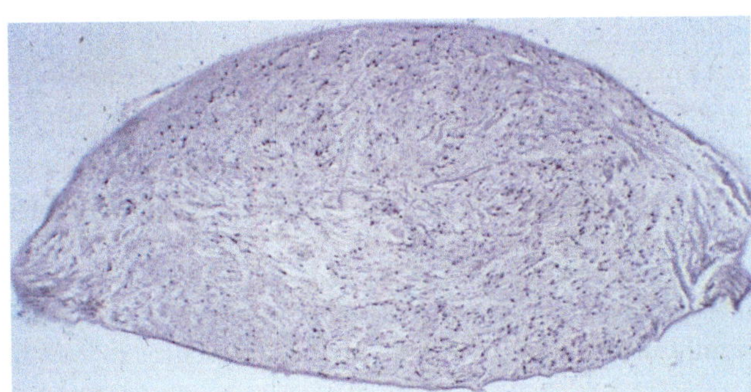

Fig. 8.3 CGRP-positive cells in canine SCG. CGRP-positive cells are diffusely distributed all over the ganglion

intralaryngeal ganglion, coexpression was confirmed only with VIP but not with CGRP. As we hypothesized that the laryngeal nerve fiber which expresses both NO and CGRP is originated from the outside of the larynx, we conducted the study to identify the coexpression of NO and CGRP in canine SCG. Immunohistochemistry for CGRP and histochemistry for NADPH diaphorase (NADPHd) were applied for a single specimen of canine SCG. The study revealed that there are a number of NADPHd-positive cells in SCG, and the major part of them expresses both NADPHd and CGRP (Fig. 8.4a, b). NADPHd-positive cells were evenly distributed in the ganglion. Of NADPHd-positive cells, 85.5 % express CGRP, and 91.5 % of CGRP-positive cells express NADPHd. These results indicate that laryngeal NANC fibers with both NO and CGRP immunoreactivity are originated from SCG.

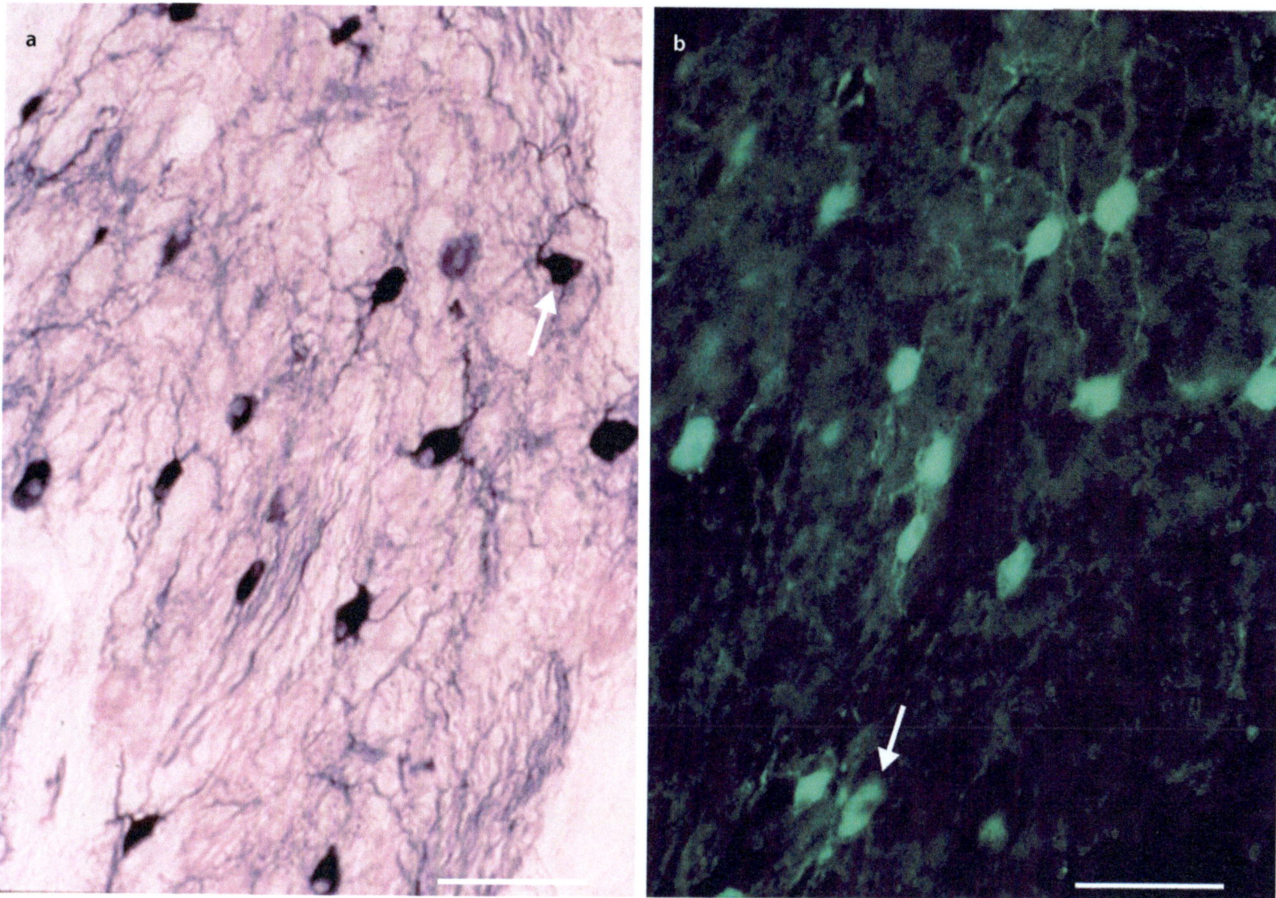

Fig. 8.4 Double staining for NADPHd (**a**) and CGRP (**b**) in canine SCG. Most of the NADPHd-positive cells also express CGRP. NADPHd(–)/CGRP(+) (**a** *arrow*) or NADPHd (–)/ CGRP(+) (**b** *arrow*) cells are rare

References

1. Li C, Horn JP. Physiological classification of sympathetic neurons in the rat superior cervical ganglion. J Neurophysiol. 2006;95:187–95.
2. Purves D, Wigston DJ. Neural units in the superior cervical ganglion of the guinea-pig. J Physiol. 1983;334:169–78.
3. Lucier GE, Egizii R, Dostrovsky JO. Projections of the internal branch of the superior cervical nerve of the cat. Brain Res Bull. 1986;16:713–21.
4. Uno T. Autonomic neurons sending fibers into the canine laryngeal nerves – using a retrograde tracer technique with cholera toxin. Nihon Jibiinkoka Gakkai Kaiho. 1993;96:66–76.
5. Hökfelt T, Johanson O, Ljugdahl Å, Lundberg JM, Schultzberg M. Peptidergic neurons. Nature. 1980;284:515–21.
6. Helke CJ, Hill KM. Immunohistochemical study of neuropeptides in vagal and glossopharyngeal afferent neurons in the rat. Neuroscience. 1998;26:539–51.
7. Kessler JA, Adler JE, Bohn MC, Black IB. Substance P in principal sympathetic neurons: regulation by impulse activity. Science. 1981;214:335–6.
8. Hökfelt T, Elfvin LG, Schultzberg M, Fuxe K, Said SI, Mutt V, Goldstein M. Immunohistochemical evidence of Vasoactive intestinal polypeptide containing neurons and nerve fibers in sympathetic ganglion. Neuroscience. 1977;2:885–96.
9. Eranko L, Paivarinta H, Soinila S, Happola O. Transmitters and modulators in the superior cervical ganglion of the rat. Med Biol. 1986;64:75–83.
10. Folan JC, Heym C. Immunohistochemical evidence for differential opioid systems in the rat superior cervical ganglion as revealed by imipramine treatment and receptor blockade. J Chem Neuroanat. 1989;2:107–18.
11. Lunberg JM, Terenius L, Hökfelt T, Martling CR, Tatemoto K, Mutt V, Polak J, Bloom S, Goldstein M. Neuropeptide Y (NPY)-like immunoreactivity in peripheral noradrenergic neurons and effects of NPY on sympathetic function. Acta Physiol Scand. 1982;116: 477–80.
12. Hisa Y. Fluorescence histochemical studies on the noradrenergic innervation of the canine larynx. Acta Anat. 1982;113:15–25.
13. Hisa Y, Matui T, Fukui K, Ibata Y, Mizukoshi O. Ultrastructural and fluorescence histochemical studies on the sympathetic innervation of the canine laryngeal glands. Acta Otolarngol. 1982;93:119–22.
14. Hisa Y, Koike S, Uno T, Tadaki N, Bamba H, Okamura H, Tanaka M, Ibata Y. Coexistence of calcitonin gene-related peptide and NADPH-diaphorase in the canine superior cervical ganglion. Neurosci Lett. 1997;228:135–8.
15. Kummer W, Heym C. Neuropeptide distribution in the cervicothoracic paraventral ganglia of the cat with particular reference to calcitonin gene-related peptide immunoreactivity. Cell Tissue Res. 1988;251: 463–71.
16. Nozaki K, Uemura Y, Okamoto S, Kikuchi H, Mizuno N. Origins and distribution of cerebrovascular nerve fibers showing calcitonin gene-related peptide-like immunoreactivity in the major cerebral artery of the dog. J Comp Neurol. 1990;297:219–26.
17. Baffi J, Gorcs T, Slowik F, Horvath M, Lekka N, Pasztor E, Palkovits M. Neuropeptides in the human superior cervical ganglion. Brain Res. 1992;570:271–8.
18. Jones MA, Marfurt CF. Calcitonin gene-related peptide and corneal innervation: a developmental study in the rat. J Comp Neurol. 1991; 313:132–50.

19. Brain SD, Williams TJ, Tippins JR, Morris HR, MacIntyre I. Calcitonin gene-related peptide is a potent vasodilator. Nature. 1985;313:54–6.
20. Goodman EC, Iversen LL. Calcitonin gene-related peptide: novel neuropeptide. Life Sci. 1986;38:2169–78.
21. Hillerdal M, Andersson SE. The effect of calcitonin gene-related peptide on the blood flow of the upper respiratory tract and the middle ear and inner ear. Acta Otolaryngol. 1989;108:94–108.
22. Lee Y, Takami K, Kawai Y, Girgis S, Hillyard CJ, MacIntyre I, Emson PC, Tohyama M. Distribution of calcitonin gene-related peptide in the rat peripheral nervous system with reference to its coexistence with substance P. Neuroscience. 1985;15:1227–37.
23. Nichols K, Krantis A, Staines W. Histochemical localization of nitric oxide-synthesizing neurons and vascular sites in the guinea-pig intestine. Neuroscience. 1992;51:791–9.
24. Hisa Y, Tadaki N, Uno T, Koike S, Tanaka M, Okamura H, Ibata Y. Nitrergic innervation of the rat larynx measured by nitric oxide synthase immunohistochemistry. Ann Otol Rhinol Laryngol. 1996;105: 550–4.
25. Hisa Y, Uno T, Tadaki N, Koike S, Banba H, Tanaka M, Okamura H, Ibata Y. Relationship of neuropeptide to nitrergic innervation of the canine laryngeal glands. Regul Pept. 1996;66:197–201.

Nodose Ganglion

Ryuichi Hirota, Hiroyuki Okano, and Yasuo Hisa

R. Hirota
Hirota ENT Clinic, Nishinoyama Nakatorii-cho, Yamashina-ku,
Kyoto 607-8306, Japan
e-mail: rhirota@koto.kpu-m.ac.jp

H. Okano • Y. Hisa (✉)
Department of Otolaryngology-Head and Neck Surgery,
Kyoto Prefectural University of Medicine, Kawaramachi-Hirokoji,
Kamigyo-ku, Kyoto 602-8566, Japan
e-mail: yhisa@koto.kpu-m.ac.jp

Y. Hisa (ed.), *Neuroanatomy and Neurophysiology of the Larynx*, DOI 10.1007/978-4-431-55750-0_9

9.1 Introduction

The vagus nerve, as its name suggests, is a nerve that follows an extremely complex course. It includes: (a) Supply to the postauricular and external auditory canal skin and afferent fibers that transmit general sensory perception from these regions (b) General visceral afferent fibers from the visceral organs in the pharynx, larynx, trachea, esophagus, thorax, and abdomen (c) Special visceral afferent fibers transmitting sensory information from the taste buds located near the epiglottis (d) General visceral efferent fibers distributed through the thoracic and abdominal viscera as parasympathetic neurons (e) Special visceral efferent fibers for arbitrary control of the striated muscle in the pharynx and larynx.

The vagus nerve, which is formed by a convergence of nerve roots arising from the lateral accessory olivary nucleus in the medulla oblongata, exits the lower surface of the skull via the jugular foramen. The superior ganglion and the inferior ganglion are also formed in and pass through this foramen. Generally, the superior ganglion is known as the jugular ganglion, whereas the inferior ganglion is known as the nodose ganglion.

General and specific visceral afferent fibers are all formed from protuberances on the pseudounipolar nerve cells found in the nodose ganglion. The central nerve fibers leaving the nodose ganglion end on the solitary tract nucleus. In addition, the sensory information from within the larynx travels from the superior and inferior laryngeal nerves, via the nodose ganglion and mainly transmitted to the interstitial subnucleus with the solitary tract nucleus in the medulla oblongata [1, 2] (see solitary tract nucleus).

9.2 Nodose Ganglion

9.2.1 Nodose Ganglion Neurons

In humans, the nodose ganglion is approximately 25 mm in length and 5 mm wide at the center, making it a fairly large spindle-shaped ganglion. Branches from the hypoglossal nerve, the superior cervical ganglion, and the first and second cervical nerves pass through this neuronal ganglion. Jones [3] conducted a measurement study on the nodose ganglion in cats and calculated the total number of neurons contained within the ganglion to be approximately 30,000. Mohiuddin [4] observed histologically that there were a small number of pseudounipolar cells mixed with the fusiform bipolar neurons. They also reported that the size of the cell body was between 35 and 40 μm and that there were also smaller cell bodies that were 20–30 μm in size.

Generally, the sensory neurons can be broadly classified into three groups based on their appearance under an electron microscope [5]. Type A cells are large and have a bright cell body. Type B cells are medium sized and are surrounded by a dense mass of perinuclear, intracellular organelles. Type C cells are small and have a Golgi apparatus close to the nucleus.

9.2.2 Neurotransmitters

With the introduction of immunohistochemistry, it has been elucidated that various neurotransmitters are present in the nodose ganglion. In 1978, Lundberg et al. [6] reported that although there were large numbers of substance P (SP)-positive cells and moderate numbers of vasoactive intestinal polypeptide (VIP)-positive cells distributed throughout the nodose ganglion, there were few cholecystokinin (CCK)-positive or somatostatin (SOM)-positive cells. In addition, they were able to elucidate the presence of calcitonin gene-related peptide (CGRP)-positive cells in the nodose ganglion, as well as the coexistence and interaction of SP and neurokinin A (NKA) [7]. In recent years, the presence of neurons possessing various neurotransmitters, such as Leu-enkephalin (ENK) and galanin (GAL), has been reported [8–10].

Nozaki et al. [11] reported that 20–30 % of all cells in the nodose ganglion are positive for neuronal nitric oxide synthase (nNOS), which is the enzyme responsible for synthesizing nitric oxide (NO), a gasotransmitter. We used NADPH diaphorase (NADPHd) histochemistry to identify the presence of NO in the nodose ganglion and investigated the concomitant presence of other neurotransmitters.

Price and Mudge [12] reported the presence of enzymes responsible for catecholamine synthesis in the sensory neuron ganglia in dorsal root ganglion. We also proved the presence of tyrosine hydroxylase (TH)-positive cells, which are one of the cells responsible for the synthesis of catecholamines, in the nodose ganglion.

9.2.3 Nociceptors

Most studies of the sensory receptors in the larynx were physiological investigations and were divided into chemoreceptors and mechanoreceptors. However, nociceptors and their relationship to the transmission of nociceptive laryngeal stimuli have recently begun to garner attention. Caterina et al. [13] elucidated the presence of cells staining positive for vanilloid receptor subtype 1 (VR1), a nociceptor, in the dorsal root ganglion and nodose ganglion. Vanilloid receptor-like protein 1 (VRL-1), which is homozygous to VR1, is also present in the nodose ganglion [14]. The $P2X_3$ gene, which is an ATP receptor gene, was cloned in 1995. $P2X_3$ receptors are present on primary afferent sensory neurons and are specifically expressed on cells that transmit nociceptive information (see sensory receptors). In 1997, Vulchanova et al. [15] verified that $P2X_3$ receptor-positive cells were present in the nodose ganglion. Acid-sensing ion channels (ASIC) are ion channels that open in response to proton stimuli. There are six reported subtypes of ASIC, but of those, ASIC 3 is known to be only expressed on small, sensory neurons [16].

We performed a detailed investigation of the involvement of these nociceptors in the neural control structure of the larynx using fluorescent neural tracers and immunohistochemistry techniques.

9.3 Nodose Ganglion Cells Projected to the Larynx

The nerve fibers originating in the nodose ganglion supply the respiratory organs via the larynx, the gastrointestinal organs via the pharynx, and the heart. They are involved in visceral perception in these organs. Reports suggest that the localization of the innervating neurons within the ganglion differs, depending on the organ. There are several pseudounipolar cells present in the nodose ganglion that supply the nerves to the arch of the aorta [17], and the fibers to the duodenum are present in cells on the caudal side of the ganglion [18].

In terms of the larynx, in 1912, Mohlant [19] used rabbits and reported that the cell bodies close to the superior pole of the nodose ganglion sent fibers via the superior larynx and controlled laryngeal perception. Lucier et al. [20] injected HRP into the internal branches of the laryngeal nerve in cats and clarified the location of the target neurons in the nodose ganglion. We reported on a detailed investigation of the cells in the nodose ganglion supplying fibers to the internal and external branches of the superior laryngeal nerve and the inferior laryngeal nerve in dogs [21–23].

9.4 Localization of the Nodose Ganglion Cells Sending Fibers to Each Laryngeal Nerve

9.4.1 Internal Branch of the Superior Laryngeal Nerve

We investigated the nodose ganglion cells supplying fibers to the internal branch of the canine superior laryngeal nerve.

We amputated the internal branch of the canine superior laryngeal nerve at the laryngeal inlet and infiltrated them with the neuronal tracer HRP. Four days later, after transcardial perfusion fixation, we excised the nodose ganglion, prepared thin sections, localized the labeled cells, and investigated the size of the labeled cells [21]. The labeled cells were present in the rostrolateral third of the ganglion (◘ Fig. 9.1). The percentage of labeled cells compared to the total number of cells was approximately 12 %, and the cell size was approximately 15–45 μm, indicating the presence of many small to medium cells. Almost no cells greater than 45 μm in size were observed (◘ Fig. 9.2). As subsequently mentioned, the percentage of

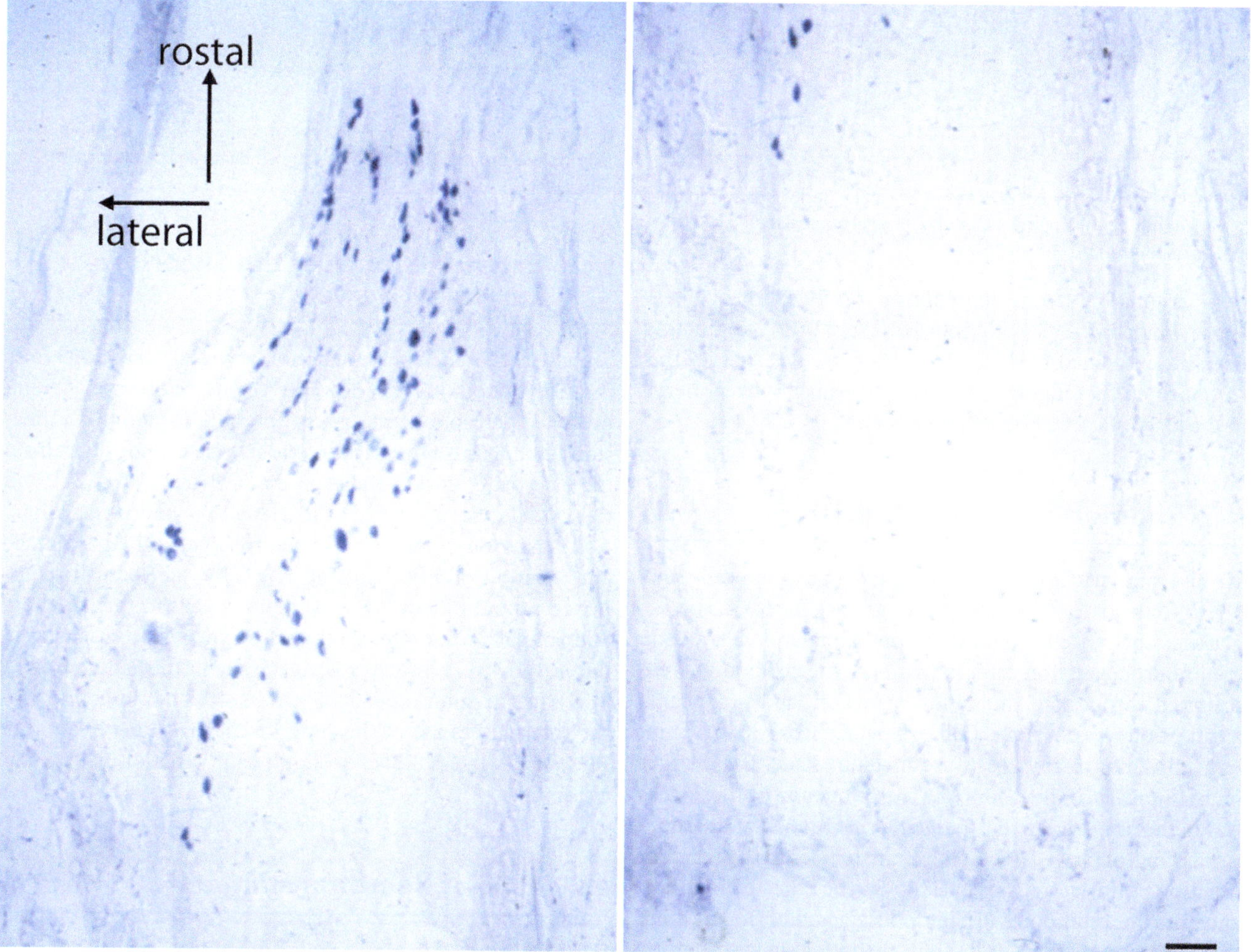

◘ Fig. 9.1 Section of the internal branch of the superior laryngeal nerve, canine nodose ganglion infiltrated with HRP. The labeled cells were present in the rostrolateral part of the ganglion [21]

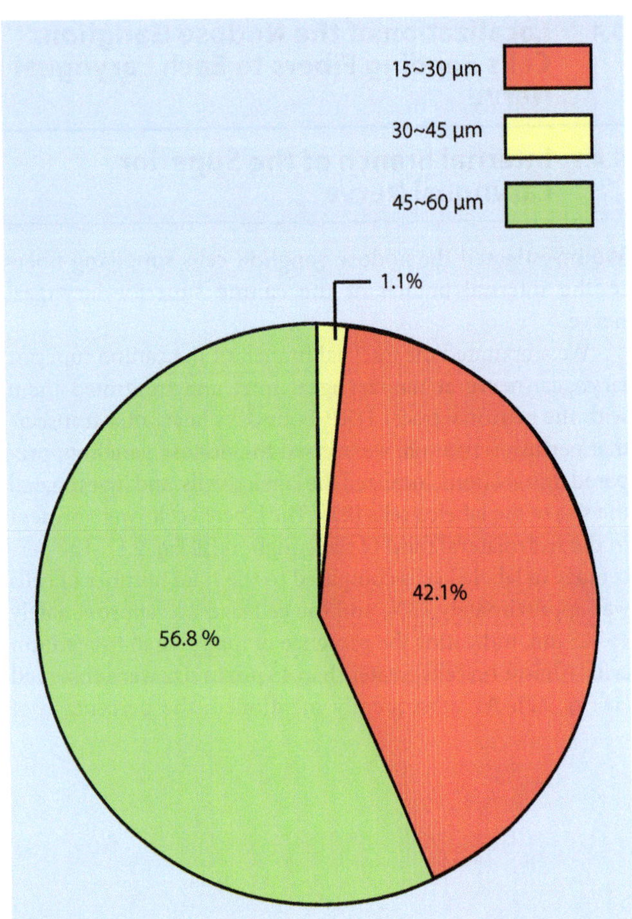

■ **Fig. 9.2** The proportion of the sizes of canine nodose ganglion cells in the internal branch of the superior laryngeal nerve, labeled by means of infiltration with HRP [21]

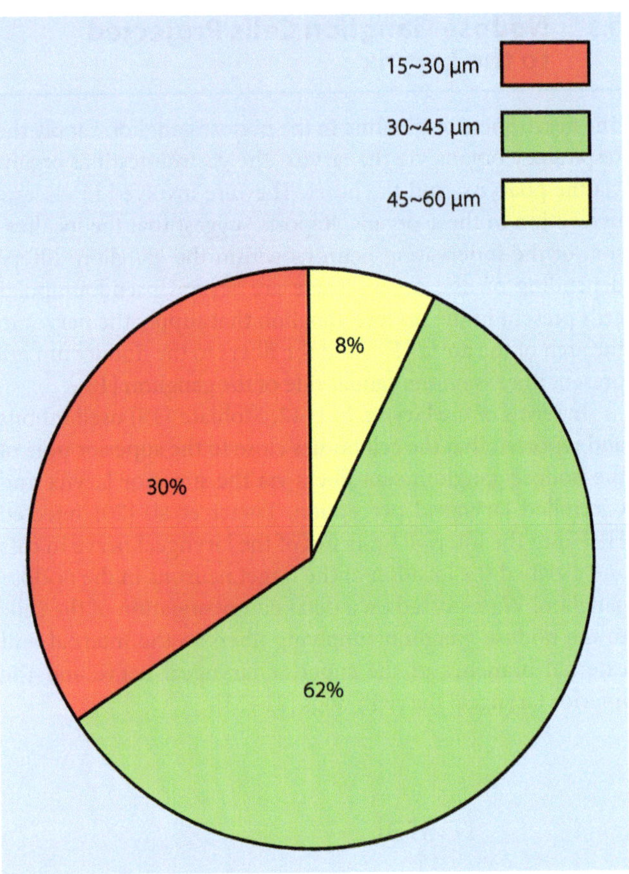

■ **Fig. 9.3** The proportion of the sizes of canine nodose ganglion cells in the inferior laryngeal nerve, labeled by means of infiltration with HRP [23]

cells originating from the inferior laryngeal nerve was approximately 0.2 %, with approximately 0.1 % originating from the external branch of the superior laryngeal nerve. This reconfirmed the importance of the superior laryngeal nerve in the laryngeal sensory nervous system.

9.4.2 Inferior Laryngeal Nerve

We also similarly investigated the nodose ganglion cells supplying fibers to the canine inferior laryngeal nerve [23].

The labeled cells comprised approximately 0.2 % of the total nodose ganglion cells. In addition, the cell bodies were scattered throughout the nodose ganglion, and no specific localization was observed. Cell size ranged from 30 to 45 μm, indicating that many cells were medium sized. Cells larger than 45 μm accounted for 8 %. The frequency of these cells was high compared to that in the internal branch of the superior laryngeal nerve (■ Fig. 9.3). If differences in the receptor function in the larynx are considered to be reflected in differences in the cell bodies in the nodose ganglion innervating them, differences such as these in the ratios suggest that the sensory receptors for the inferior laryngeal nerve below the glottis may play a different role to those of the internal branch of the superior laryngeal nerve.

9.4.3 External Branch of the Superior Laryngeal Nerve

The external branch of the superior laryngeal nerve innervates the cricothyroid muscle; however, the nerve fibers that result in innate perception and perception of the anterior commissure submucosa of the cricothyroid muscle include fibers from the external branch of the superior laryngeal nerve [24], and these cell bodies appear to be located in the nodose ganglion.

As previously mentioned, we investigated the percentage and location of all cells using HRP. The labeled cells comprised no more than 0.1 % of the total cells in the ganglion, and similar to the internal branch of the superior laryngeal nerve, they were located rostrolaterally in the ganglion [22]. The largest region was 30–45 μm in size and, in contrast to the internal branch of the superior laryngeal nerve and the inferior laryngeal nerve, no significant variation in size was observed.

9.5 Role of Neurotransmitters

9.5.1 CGRP

We investigated the presence of CGRP-positive cells in the canine nodose ganglion.

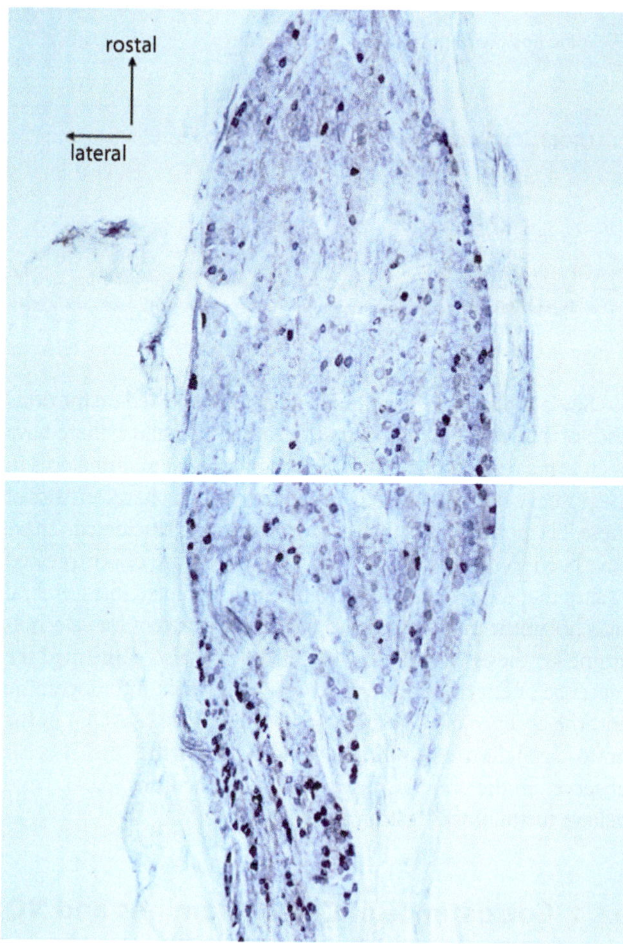

Fig. 9.4 The CGRP-positive cells are commonly distributed rostrolaterally in the canine nodose ganglion, but no clear pattern of localization was observed in any of the sections

Our measurements indicated that the number of cells in the nodose ganglion was approximately 30,000, which is consistent with the results of previous reports. There were approximately 7200 CGRP-positive cells comprising approximately 24 % of the total. All cells were 30 μm or larger, clarifying that medium to large cells are CGRP positive. The large cells in particular, i.e., those greater than 45 μm in size, comprised 95 % of all CGRP-positive cells. The CGRP-positive cells were diffusely present and within the nodose ganglion, there was no obvious localization (Fig. 9.4). In addition, cells with coexistence of CGRP and SP were observed. We believe that the coexistence of CGRP and SP may potentiate the actions of SP [25].

9.5.2 NO

We used NADPHd histochemistry to investigate the NO affinity of cells in the nodose ganglion in dogs, rats, and guinea pigs [26].

We observed three types of cells, namely, those that were densely stained, those that were weakly stained, and those that were barely stained. The percentage of densely stained cells in rats was approximately 25 %, and including the weakly stained cells, approximately 90 % were NADPHd positive. In dogs, the percentage of NADPHd-positive cells was somewhat high at approximately 95 %, and the staining pattern in guinea pigs showed virtually no difference to that in rats. The distribution of NADPHd-positive cells in the nodose ganglion was generally diffuse and spread from rostral to caudal in rats, while medium to large cells were most common in terms of size (Fig. 9.5). However, there were very few densely stained cells observed rostrally with NADPHd histochemistry in dogs.

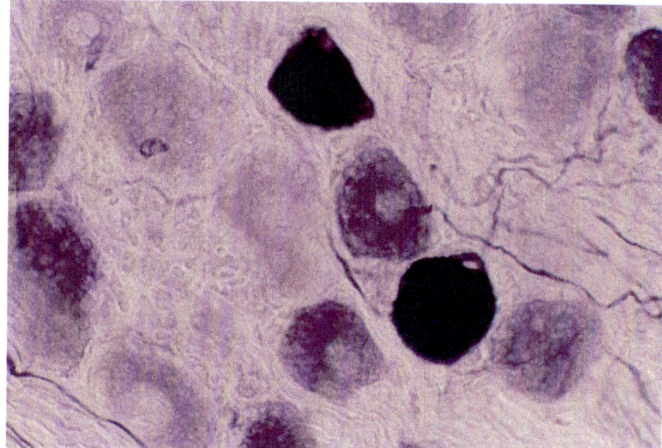

Fig. 9.5 Three types of cells that stained positive for NADPHd were observed in dogs: cells that were densely stained, cells that were weakly stained, and cells that were barely stained

The neurons that were weakly stained during our NADPHd histochemistry tests also stained positively during nNOS immunohistochemistry, and we believe they possess nNOS. Accordingly, although approximately 30 % of rat nodose ganglion cells have been reported to be NADPHd positive to date [27], we believe that 90 % of the nodose ganglion cells in the three types found during this investigation stain positive for nNOS. We indicate that specific differences in the proportion of neuronal cells with nNOS are present in the superior cervical ganglion [28]. Based on the results of this study, the percentage of cells that stain positive for NADPHd during histochemistry in the nodose ganglion differ based on their type, and it appears that there are differences in NO amounts even in primary sensory neurons, where NO participates in the transmission of information. The nodose ganglion neurons that send fibers to the internal branch of the superior laryngeal nerve are located rostrally in the ganglion [28]. Because there are few cells in the rostral area that are densely stained during NADPHd histochemistry in this area in dogs, the cells that are chiefly involved in laryngeal sensation may be the cells that are weakly stained during NADPHd histochemistry. However, the function of the neurons and the relation to NADPHd histochemistry staining has not yet been clarified.

9.5.3 Coexistence of NO and CGRP

We performed fluorescent immunohistochemistry of CGRP and NADPHd histochemistry on the same sections and investigated the coexistence of NO and CGRP in dogs.

Table 9.1 The percentage of coexistence of CGRP and NADPHd in the canine nodose ganglion

$\frac{\text{NADPHd}(+)\cdot\text{CGRP}(+)}{\text{CGRP}(+)} = 93.1 \pm 1.6\%$	$\frac{\text{NADPHd}(++)\cdot\text{CGRP}(+)}{\text{CGRP}(+)} = 6.3 \pm 0.9\%$
$\frac{\text{NADPHd}(+)\cdot\text{CGRP}(+)}{\text{NADPHd}(+)} = 28.6 \pm 2.3\%$	$\frac{\text{NADPHd}(++)\cdot\text{CGRP}(+)}{\text{NADPHd}(++)} = 20.4 \pm 2.8\%$

The percentage of cells that were CGRP positive that were also densely stained during NADPHd histochemistry was approximately 6.3 %, while, conversely, approximately 20.4 % of the cells that were densely stained during NADPHd histochemistry were also CGRP positive (Table 9.1).

In addition, a similar value of 28.6 % was obtained when we examined the percentage of cells that were weakly stained during NADPHd histochemistry and were coexistent with CGRP-positive cells. However, because these percentages were roughly the same in each individual, we believe that the there are differences in the cell groups stained during NADPHd histochemistry and that there are also difference in their functions. In the rat nodose ganglion, it is known that almost all CGRP is present in almost all SP-positive cells. Conversely, SP is present in approximately one third of CGRP-positive cells [29], indicating that CGRP-positive cells can be divided into at least two groups. Based on these results, the CGRP-positive cells can be further subdivided based on their staining when exposed to NADPHd and suggest that there may be various, functionally different groups of cells.

9.5.4 Catecholamine-Containing Cells

We used TH immunohistochemistry, which is one way to study catecholamine synthesis, and investigated the presence of catecholamine-containing cells within the nodose ganglions of dogs [30] and rats [31].

Of all the cells within the nodose ganglion, many TH-positive cells (approximately 2.5–8.0 %) were observed in the rostrolateral to central regions. The distribution within the nodose ganglion showed that, compared to other sites, there was a comparatively high distribution of densely staining TH-positive cells and small cells in the rostrolateral region (Fig. 9.6).

Next, we examined the small cells and densely TH-positive cells that were observed in the rostrolateral region using an electron microscope.

The substances that were TH immunopositive were observed as dark deposits scattered diffusely throughout the cytoplasm. Cytoplasmic organelles such as mitochondria and rough endoplasmic reticulum have developed, and these cells are believed to correspond to type B cells (Fig. 9.7).

Since 1983, when Price and Mudge [12] reported on the presence of TH-positive cells in the dorsal root ganglion, there have been some reports [32–34] on catecholamine-containing cells in the primary sensory ganglia. However, there are characteristics of these TH-positive cells that have not yet been elucidated. There have been some reports investigating TH-positive cells expressed in sites that differ from the conventionally accepted intracerebral catecholamine distribution and the lack of enzymes for catecholamine synthesis other than TH [35, 36]. We also confirmed the presence of enzymes that would convert L-dopa into dopamine and the absence of L-amino acid decarboxylase (AADC) in the nodose ganglion. Accordingly, we wonder whether the TH cells observed in the present study were L-dopa neurons [35–37] and believe further investigation is required.

9.5.5 Coexistence of Catecholamines and NO

We investigated the coexistence of NO and catecholamines in the nodose ganglion using TH immunohistochemistry and NADPHd histochemistry [31].

Of the TH-positive cells, 54.8 % were noted to be NADPHd positive, but only 19.9 % of the NADPHd-positive cells were noted to be TH positive. In addition, 72.2 % of the nodose ganglion cells that projected fibers to the solitary tract nucleus were positive for both NADPHd and TH.

Next, we investigated the role played by TH-positive cells in the sensory innervation of the larynx in the nodose ganglion in dogs. The nodose ganglion cells, which send fibers to the internal branch of the superior laryngeal nerve, were labeled with gold-labeled cholera toxin (CTBG), and we performed TH immunohistochemistry [30].

Of the labeled cells in the nodose ganglion, two to three cells were noted to be TH positive. We also investigated the central projections of these TH-positive cells. After injecting CTBG into the solitary tract nucleus, we excised the nodose ganglion on the injected side and also performed TH immunohistochemistry. The results showed that all TH-positive cells were also labeled with CTBG.

The results also indicated that all TH-positive cells have central endings, terminating in the solitary tract nucleus, and some of the TH-positive cells were elucidated to send neuronal fibers to the larynx. Accordingly, the TH-positive cells that send these fibers to the larynx were verified to be involved in sensory transmission from the larynx.

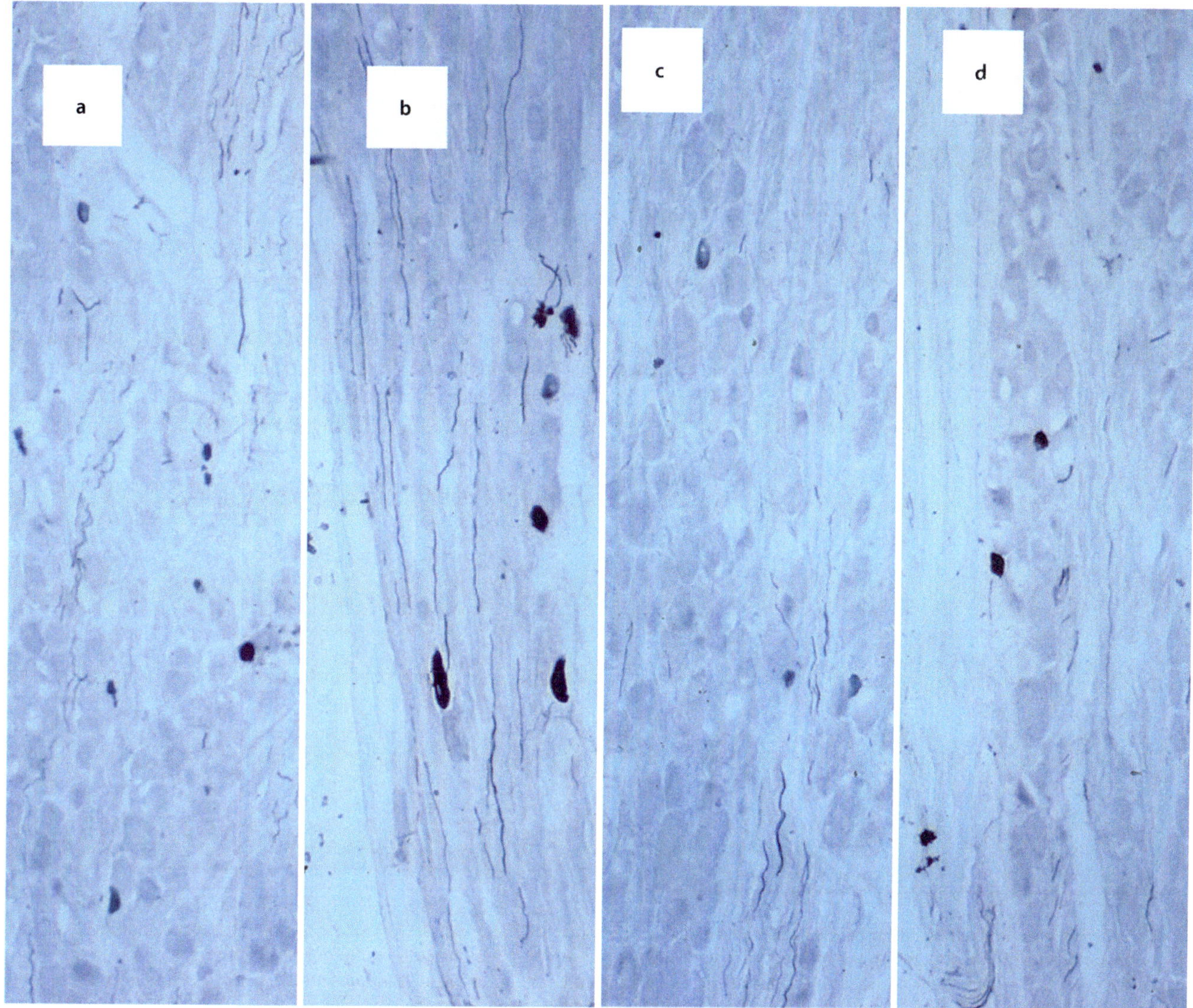

Fig. 9.6 TH-positive cells in the canine nodose ganglion (**a** rostrolateral, **b** caudolateral, **c** rostromedial, **d** caudomedial). Compared to the other regions, the rostrolateral area has a greater distribution of small cells and cells that stained densely TH positive [30]

9.6 Role of Nociceptors

9.6.1 Capsaicin Receptors

(a) VR1 and VRL-1 in the nodose ganglion

We investigated VR1 and VRL-1 in the nodose ganglion of rats using immunohistochemistry.

VR1 was expressed in comparatively small to medium cells and VRL-1 was expressed in comparatively moderate to large cells. Of all cells, approximately 50 % were shown to be VR1 positive (Fig. 9.8a), and approximately 11 % of cells were VRL-1 positive [38] (Fig. 9.8b). In addition, we noted expression of both VR1 and VRL-1 in moderately sized cells using the fluorescent double staining method, and approximately 60 % of the VRL-1-positive cells were VR1 positive (Fig. 9.9).

VR1 is mainly expressed in small C cells that supply the non-medullary fibers, and VRL-1 is mainly expressed on large Aδ cells that supply the medullary fibers [39, 40]. Similar trends are also observed in the nodose ganglion.

To date, there have been no reports regarding the coexistence of VR1/VRL-1 in the dorsal root ganglion or trigeminal ganglion. There may be results regarding the multitude of coexistent cells in the nodose ganglion and the differences and relationship between the innervation regions of each of the sensory neuron ganglia. This topic is an issue that requires investigation in the future.

(b) The role of VR1 and VRL-1 in the laryngeal nervous system

We investigated the role of the VR1-positive and VRL-1-positive cells that are present in the nodose ganglion in laryngeal sensory innervation. We exposed the internal branch of the superior laryngeal nerve in rats and injected Fluoro-Gold (FG) as a neuronal marker. After 3 days of transcardiac perfusion fixation, we excised the nodose ganglion. After preparing frozen sections, we performed fluorescent immunohistochemistry for both VR1 and VRL-1.

Of the FG-labeled cells, the percentage of VR1-positive cells was 49.0 ± 4.4 % (Fig. 9.10a), and the percentage of

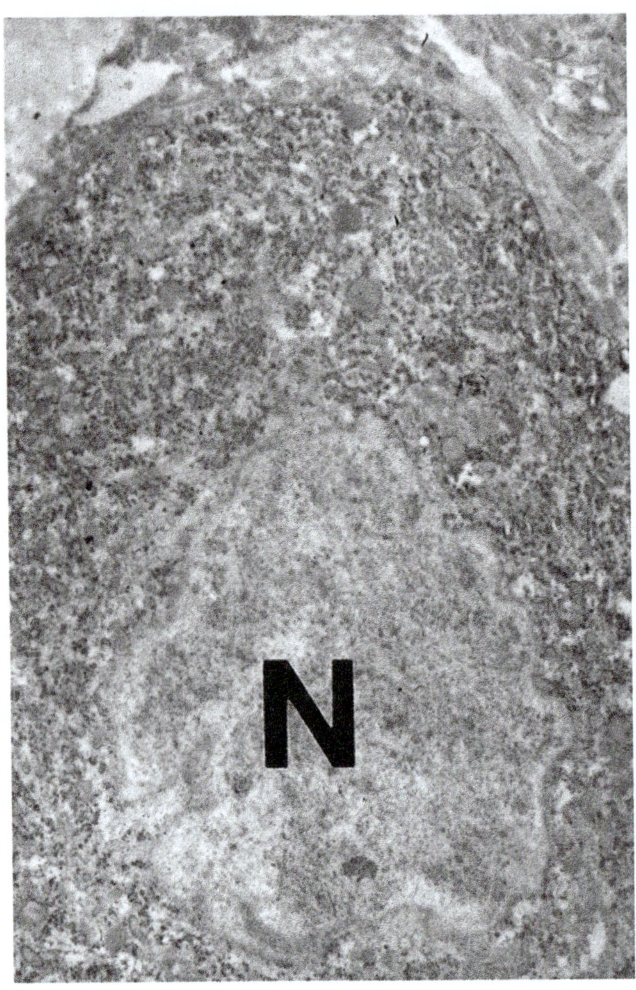

■ **Fig. 9.7** TH-immunopositive cells were observed to have dark deposits scattered diffusely throughout the cytoplasm. Cytoplasmic organelles such as the mitochondria and rough endoplasmic reticulum have developed [30]

VRL-1-positive cells was 12.5±4.1% (■ Fig. 9.10b). The percentages were roughly the same as the percentage of VR1 and VRL-1-postiive cells. Based on the above, both VR1 and VRL-1 play a role in the transmission of laryngeal nociceptive stimuli, and VR1 is presumed to play an important role.

9.6.2 ATP Receptors

(a) P2X3 and in the nodose ganglion

ATP receptors are largely divided into the P2X family of ion channels and the P2Y family that undergo G-protein coupling. A further classification into more than seven subtypes has been reported. Of these, P2X3 receptors are specifically expressed on primary sensory

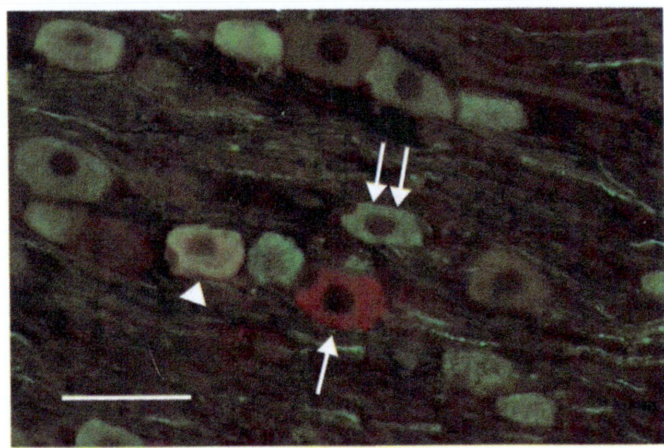

■ **Fig. 9.9** Coexistence of VR1- and VRL-1-positive cells in the rat nodose ganglion. Of the VRL-1-positive cells, approximately 60% were VR1 positive. VRL-1-only-positive cells (*single arrow*), VR1-only-positive cells (*double arrow*), coexistent VR1 and VRL-1 (*arrow head*) [38]

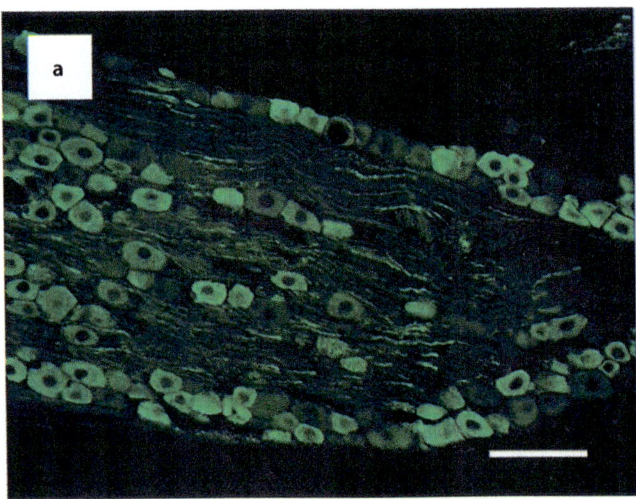

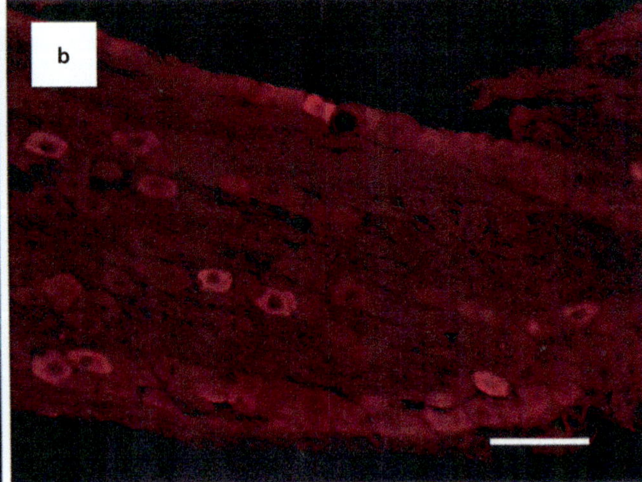

■ **Fig. 9.8** VR1-positive cells (**a**) and VRL-1-positive cells (**b**) in the rat nodose ganglion. Approximately 50% of all cells are positive for VR1, which is expressed in comparatively small- to medium-sized cells. Approximately 10% of all cells are positive for VRL-1, which is expressed in comparatively moderately sized to large cells [38]

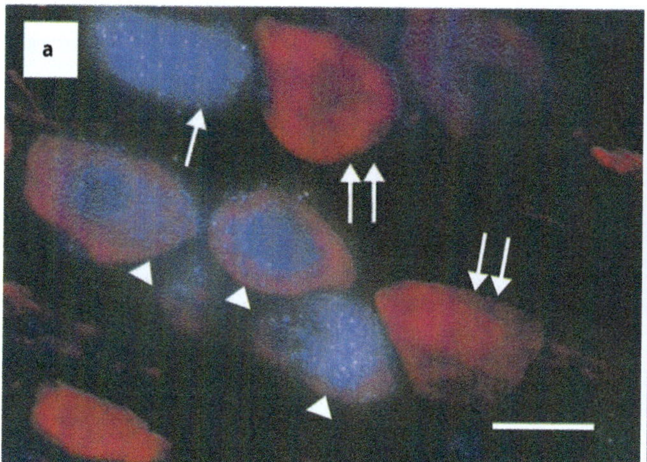

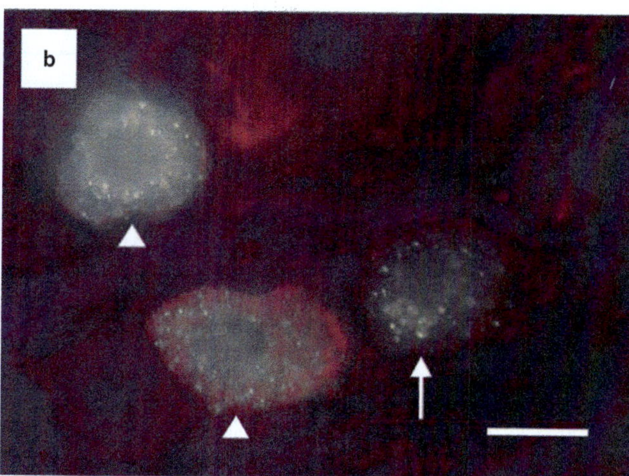

Fig. 9.10 VRl-positive cells sending laryngeal fibers in rats. (**a**) Of the FG-labeled cells, 49 % were positive for VR1. FG-labeled cell (*single arrow*), VR-1-positive cells (*double arrow*), VR1-positive FG-labeled cell (*arrow head*). (**b**) Of the FG-labeled cells, 12.5 % were positive for VRL-1. FG-labeled cell (*single arrow*), VRL-1-positive FG-labeled cell (*arrow head*)

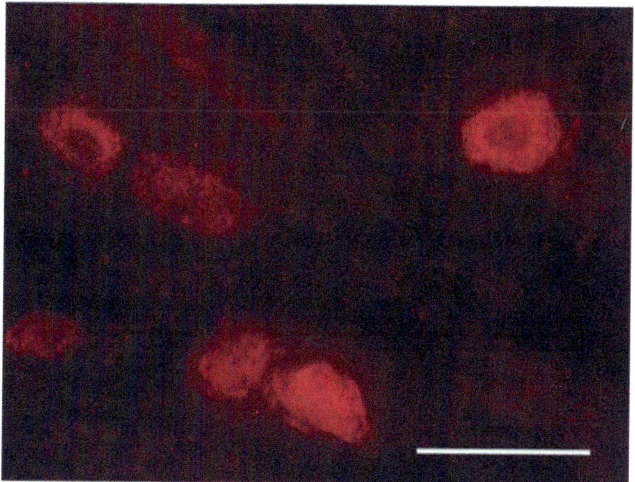

Fig. 9.11 $P2X_3$-positive cells in the rat nodose ganglion. Of all cells, 21.2 % were $P2X_3$ positive

neurons and are reported to be related to nociception and the bladder capacity reflex [41]. Of all the cells in the nodose ganglion, 21.2 % were noted to be P2X3-positive cells (Fig. 9.11). There was no clear localization pattern within the nodose ganglion. P2X3 cells are reported to comprise 32.7 % of the dorsal root ganglion and 26.7 % of the trigeminal ganglion [42], and the rate of positive cells in the nodose ganglions that we studied showed comparatively lower results than other ganglions. This could reflect the fact that the dorsal root ganglion and trigeminal ganglion are formed from neurons that send general afferent fibers, and compared to them, the nodose ganglion is made up of neurons that send general and specific visceral afferent fibers.

(b) The role of P2X3 in the laryngeal nervous system

We used the abovementioned test with FG to investigate the role of P2X3-positive cells located in the nodose ganglion in the laryngeal nervous system.

Of the cells that were FG labeled, 36.7 % were P2X3 positive. The nodose ganglion cells are cell bodies with general and specific visceral afferent fibers; however, there are also differences in the nociceptive stimulus receptors in the visceral afferent fibers from thoracic and gastrointestinal organs, such as those from the pharynx, larynx, and trachea. Differences in the rate of P2X3 positivity may reflect differences in the nociceptive stimuli receptors. In addition, there are said to be interactions between P2X3 and SP. By investigating the relationship to SP in the nodose ganglion and to the submucosal expression of SP on the laryngeal surface of the epiglottis, we believe we may be able to elucidate the pathology of a cough of unknown origin.

9.6.3 Acid-Sensitive Receptors

In recent years, the relationship between abnormal sensation in the pharynx and gastroesophageal reflux disease has been receiving attention. One of the causes for this is believed to be a direct action of gastric acid on the laryngopharyngeal area. Acid-sensitivity receptors may play a role in the mechanism of onset of pain and the abnormal sensation caused by acid stimuli. In order to investigate the relationship between the laryngeal sensory nervous system and the acid-sensitivity receptors, we investigated the expression of ASIC3 in the nodose ganglion. ASIC3, which is one of the receptors in the acid-sensitive receptor family and also the only one that is specifically expressed in sensory nerves, was examined using immunohistochemistry.

Of all the cells in the nodose ganglion, 11.4 % were ASIC3 positive. Going forward, we will create animal models for the administration of acidic or inflammatory substances. By studying the changes in the ASIC3-positive cells in the nodose ganglion and the larynx, we hope to investigate the origin of the sensory abnormalities in the larynx.

References

1. Kalia M, Mesulam MM. Brain stem projections of sensory and motor components of the vagus complex in the cat: II. Laryngeal, tracheobronchial, pulmonary, cardiac, and gastrointestinal branches. J Comp Neurol. 1980;193:467–508.
2. Hisa Y, Lyon MJ, Malmgren LT. Central projection of the sensory component of the rat recurrent laryngeal nerve. Neurosci Lett. 1985;55:185–90.
3. Jones RL. Cell fibre ratios in the vagus nerve. J Comp Neurol. 1937;67: 469–82.
4. Mohiuddin A. Vagal preganglionic fibres to the alimentary canal. J Comp Neurol. 1953;99:289–317.
5. Lieberman AR. Sensory ganglia. In: Landon DN, editor. The peripheral nerve. New York: Wiley; 1976. p. 188–278.
6. Lundberg JM, Hökfelt T, Nilsson G, Terenius L, Rehfeld J, Elde R, Said S. Peptide neurons in the vagus, splanchnic and sciatic nerves. Acta Physiol Scand. 1978;104:499–501.
7. Lundberg JM, Franco-Cereceda A, Hua X, Hökfelt T, Fischer JA. Coexistence of substance P and calcitonin gene-related peptidelike immunoreactivities in sensory nerves in relation to cardiovascular and bronchoconstrictor effects of capsaicin. Eur J Pharmacol. 1985;108: 315–9.
8. Helke CJ, Hill KM. Immunohistochemical study of neuropeptides in vagal and glossopharyngeal afferent neurons in the rat. Neuroscience. 1988;26:539–51.
9. Philippe C, Cuber JC, Bosshard A, Rampin O, Laplace JP, Chayvialle JA. Galanin in porcine vagal sensory nerves: immunohistochemical and immunochemical study. Peptides. 1990;11:989–93.
10. Hisa Y, Tadaki N, Uno T, Okamura H, Taguchi J, Ibata Y. Neuropeptide participation in canine laryngeal sensory innervation. Immunohistochemistry and retrograde labeling. Ann Otol Laryngol. 1994;103:767–70.
11. Nozaki K, Moskowitz MA, Maynard KI, Koketsu N, Dawson TM, Bredt DS, Snyder SH. Possible origins and distribution of immunoreactive nitric oxide synthase-containing nerve fibers in cerebral arteries. J Cereb Blood Flow Metab. 1993;13:70–9.
12. Price J, Mudge AW. A subpopulation of rat dorsal root ganglion neurones is catecholaminergic. Nature. 1983;301:241–3.
13. Caterina MJ, Schumacher MA, Tominaga M, Rosen TA, Levine JD, Julius D. The capsaicin receptor: a heat-activated ion channel in the pain pathway. Nature. 1997;389:816–24.
14. Ichikawa H, Sugimoto T. Co-expression of VRL-1 and calbindin D-28k in the rat sensory ganglia. Brain Res. 2002;924:109–12.
15. Vulchanova L, Riedl MS, Shuster SJ, Buell G, Surprenant A, North RA, Elde R. Immunohistochemical study of the P2X2 and P2X3 receptor subunits in rat and monkey sensory neurons and their central terminals. Neuropharmacology. 1997;36:1229–42.
16. Chen CC, England S, Akopian AN, Wood JN. A sensory neuron-specific, proton-gated ion channel. Proc Natl Acad Sci U S A. 1998;18:10240–5.
17. Portalier P, Vigier D. Localization of aortic cells in the nodose ganglion by HRP retrograde transport in the cat. Neurosci Lett. 1979;11:7–11.
18. Filippova LV, Bagaev VA, Makarov FH, Rybakov VL. The localization of neurons in the nodose ganglion of the cat that innervate the rostral portion of the duodenum. Morfologiia. 1992;102:25–30.
19. Mohlant M. Le nerf vague: etude anatomique et expérimentale. II. Innervation motrice du larynx. Le Nevraxe. 1912;13:22–44.
20. Lucier GE, Egizii R, Dostrovsky JO. Projections of the internal branch of the superior laryngeal nerve of the cat. Brain Res Bull. 1986;16:713–21.
21. Hisa Y, Toyoda K, Uno T, Murakami Y, Ibata Y. Localization of the sensory neurons in the canine nodose ganglion sending fibers into the internal branch of the superior laryngeal nerve. Eur Arch Otorhinolaryngol. 1991;248:265–7.
22. Toyoda K. Localization of sensory neurons in the canine nodose ganglion sending fibers to the laryngeal nerves. Nihon Jibiinkoka Gakkai Kaiho. 1991;94:1888–97.
23. Toyoda K, Hisa Y, Uno T, Tadaki N. Distribution of the afferent neurons from the canine recurrent laryngeal nerve. Eur Arch Otorhinolaryngol. 1993;249:485–7.
24. Suzuki M, Kirchner JA. Afferent nerve fibers in the external branch of the superior laryngeal nerve in the cat. Ann Otol Rhinol Laryngol. 1968;77:1059–70.
25. Goodman EC, Iversen LL. Calcitonin gene-related peptide; novel neuropeptide. Life Sci. 1986;38:216–2178.
26. Koike S, Hisa Y, Uno T, Murakami Y, Tamada Y, Ibata Y. Nitric oxide synthase and NADPHdiaphorase in neurons of the rat, dog and guinea pig nodose ganglia. Acta Otolaryngol Suppl. 1998;539:110–2.
27. Aimi Y, Fujimura M, Vincent SR, Kimura H. Localization of NADPH-diaphorase-containing neurons in sensory ganglia of the rat. J Comp Neurol. 1991;306:382–92.
28. Hisa Y, Uno T, Tadaki N, Umehara K, Okamura H, Ibata Y. NADPH-diaphorase and nitric oxide synthase in the canine superior cervical ganglion. Cell Tissue Res. 1995;279:629–31.
29. Helke CJ, Niederer AJ. Studies on the coexistence of substance P with other putative transmitters in the nodose and petrosal ganglia. Synapse. 1990;5:144–51.
30. Uno T, Hisa Y, Tadaki N, Okamura H, Ibata Y. Tyrosine hydroxylase-immunoreactive cells in the nodose ganglion for the canine larynx. Neuroreport. 1996;7:1373–6.
31. Nishiyama K, Yagita K, Yamaguchi S, Kitamura S, Matsuo T, Uno T, Tanaka M, Hisa Y, Ibata Y, Okamura H. Tyrosine hydroxylase and NADPH-diaphorase in the rat nodose ganglion: colocalization and central projection. Acta Histchem Cytochem. 2001;34:135–41.
32. Katz DM, Markey KA, Goldstein M, Black IB. Expression of catecholaminergic characteristics by primary sensory neurons in the normal adult rat in vivo. Proc Natl Acad Sci U S A. 1983;80:3526–30.
33. Katz DM, Black IB. Expression and regulation of catecholaminergic traits in primary sensory neurons: relationship to target innervation in vivo. J Neurosci. 1986;6:983–9.
34. Katz DM, Adler JE, Black IB. Catecholaminergic primary sensory neurons: autonomic targets and mechanisms of transmitter regulation. Fed Proc. 1987;46:24–9.
35. Okamura H, Kitahama K, Mons N, Ibata Y, Jouvet M, Geffard M. L-dopa-immunoreactive neurons in the rat hypothalamic tuberal region. Neurosci Lett. 1988;95:42–6.
36. Okamura H, Kitahama K, Mons N, Matsumoto Y, Ibata Y, Geffard M. Heterogeneous distribution of L-DOPA immunoreactivity in dopaminergic neurons of the rat midbrain. In: Nagatsu T, editor. Basic, clinical, and therapeutic aspects of Alzheimer's and Parkinson's diseases, vol. 1. New York: Plenum Press; 1990. p. 423–6.
37. Misu Y, Goshima Y, Ueda H, Okamura H. Neurobiology of L-DOPAergic systems. Prog Neurobiol. 1996;49:415–54.
38. Uno T, Koike T, Bamba H, Hirota R, Hisa Y. Capsaicin receptor expression in the rat laryngeal innervation. Ann Otol Rhinol Laryngol. 2004;113:356–8.
39. Tominaga M, Caterina MJ, Malmberg AB, Rosen TA, Gilbert H, Skinner K, Raumann BE, Basbaum AI, Julius D. The cloned capsaicin receptor integrates multiple pain-producing stimuli. Neuron. 1998;21:531–43.
40. Tominaga M. Itami Juyoutai Kenkyu no Shinpo. Nou no kagaku (Brain Sci). 2001;23:829–35.
41. Cockayne DA, Hamilton SG, Zhu QM, Dunn PM, Zhong Y, Novakovic S, Malmberg AB, Cain G, Berson A, Kassotakis L, Hedley L, Lachnit WG, Burnstock G, McMahon SB, Ford AP. Urinary bladder hyporeflexia and reduced pain-related behaviour in P2X3-deficient mice. Nature. 2000;407:1011–5.
42. Tsuzuki K, Kondo E, Fukuoka T, Yi D, Tsujino H, Sakagami M, Noguchi K. Differential regulation of P2X3mRNA expression by peripheral nerve injury in intact and injured neurons in the rat sensory ganglia. Pain. 2001;91:351–60.

Projections to the Brain Stem

Nucleus Ambiguus

Shigeyuki Mukudai, Yoichiro Sugiyama, and Yasuo Hisa

S. Mukudai
Department of Otolaryngology-Head and Neck Surgery, Kyoto Prefectural University of Medicine, Kawaramachi-Hirokoji, Kamigyo-ku, Kyoto 602-8566, Japan
e-mail: s-muku@koto.kpu-m.ac.jp

Y. Sugiyama • Y. Hisa (✉)
Department of Otolaryngology-Head and Neck Surgery, Kyoto Prefectural University of Medicine, Kawaramachi-Hirokoji, Kamigyo-ku, Kyoto 602-8566, Japan
e-mail: yhisa@koto.kpu-m.ac.jp

Y. Hisa (ed.), *Neuroanatomy and Neurophysiology of the Larynx*, DOI 10.1007/978-4-431-55750-0_10

10.1 Anatomical Organization of the Brainstem Nuclei That Regulate the Laryngeal Motor Activity

The brainstem consisted of the midbrain, pons, and medulla oblongata plays a significant role in the regulation of laryngeal functions. For example, the nucleus tractus solitarius (NTS) in the medulla oblongata receives input from various types of visceral afferents including laryngeal, pharyngeal, and pulmonary afferents, which contributes to the regulation of breathing, phonation, and the airway protective reflexes including swallowing and coughing. On the other hand, the motoneurons that project to the pharyngeal, laryngeal, and esophageal musculatures in the nucleus ambiguus (NA) produce motor sequence of respiratory and non-respiratory behaviors. In addition, the efferent from the dorsal motor nucleus of vagus (DMNV) involved in the parasympathetic autonomic regulation provides not only esophageal peristalses but also secretion of the larynx, which may contribute to the regulation of these behaviors. We thus focused on these medullary nuclei that could constitute the neural networks that control sensory, motor, and autonomic activity of the larynx. As such, we revealed the efferent and afferent projections of the larynx using retrograde or anterograde tracer injection to the specific region of the larynx.

10.2 Nucleus Ambiguus

The NA is consisted of the rostrocaudally extended column from the level of the obex to the caudal portion of the retrofacial nucleus in the ventrolateral medulla, which is subdivided by three subnuclei regarding density of the neurons: compact formation (NAc), semicompact formation (NAsc), and loose formation (NAl). Previous studies have indicated the specific locations of motoneurons in the NA that innervate to the specific musculatures of the pharynx, larynx, and esophagus. For example, neurons that innervate to the esophageal muscles are located in the NAc, whereas the pharyngeal motoneurons are mainly distributed in the NAsc. Meanwhile, laryngeal motoneurons other than those that project to the cricothyroid muscle are distributed in the NAl. ◘ Figure 10.1 indicates the location of neurons in the NA that innervate to the pharyngeal, laryngeal, and esophageal muscles in felines [1–4, 28].

While many investigators have shown the locations of pharyngeal, laryngeal, and esophageal motoneurons in the NA using a retrograde neuronal tracer, we identified the relative localization among the laryngeal motoneurons that innervate the intrinsic laryngeal muscles including the thyroarytenoid (TA), posterior cricothyroid (PCA), lateral cricoarytenoid (LCA), arytenoid (Ary), and cricothyroid (CT) muscles using a multi-tracer study [5, 6].

Previous studies have noted that neurons in the NA possess the neurotransmitter including acetylcholine, glutamate, galanin, and calcitonin gene-related peptide (CGRP) [7–11]. These neurotransmitters that could exist in the laryngeal motoneurons probably regulate laryngeal motor activity. Indeed, microinjection of excitatory amino acid to the NA can produce pronounced activation of the recurrent laryngeal nerve (RLN), while application of γ-aminobutyric acid (GABA) decreases the RLN activity by inhibiting the laryngeal motoneurons [12–14]. On the other hand, as reported by King et al. [15], injection of the serotonin agonists in the

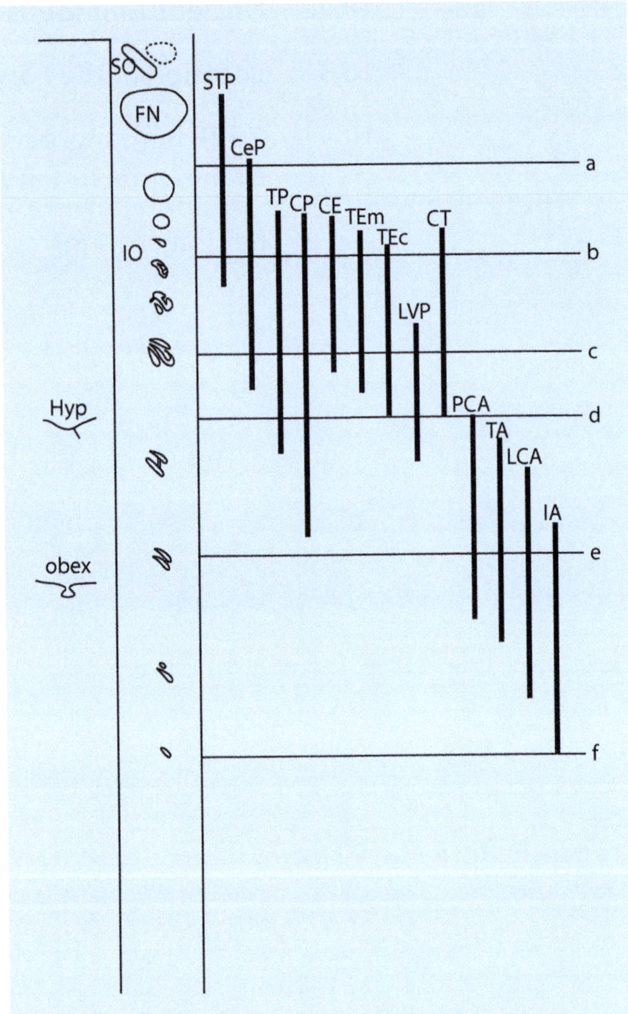

◘ **Fig. 10.1** A diagram of the feline nucleus ambiguus which demonstrates schematically the level of the labeled cell column for the pharyngeal, cervical esophagus, and laryngeal muscles in the rostrocaudal direction. The level in the brainstem is indicated with the shape of the facial nucleus (FN) and the inferior olivary nucleus (IO). (**a**) The level of the center in the facial nucleus. (**b**) The level of the rostral part in the inferior olivary nucleus. (**c**) The level where the principal nucleus of IO develops well. (**d**) The level of the rostral part in the hypoglossal nucleus. (**e**) The level of the slightly rostal to the obex. (**f**) The level of the caudal end in the inferior olivary nucleus. *CE* cervical esophagus muscle; *CeP* cephalopharyngeal muscle; *CP* cricopharyngeal muscle; *CT* cricothyroid muscle; *IA* interarytenoid muscle; *LCA* lateral cricoarytenoid muscle; *LVP* levator veli palatini muscle; *PCA* posterior cricoarytenoid muscle; *STP* stylopharyngeal muscle; *TA* thyroarytenoid muscle; *TEc* thoracic esophagus muscle, caudal portion; *TEm* thoracic esophagus muscle, middle portion; *TP* thyropharyngeal muscle vicinity of the NA attenuates the RLN activity, which suggests that serotonin may act as an inhibitory neurotransmitter of the laryngeal motoneurons. We ascertained whether there were significant changes in immunoreactivity of CGRP in the laryngeal motoneurons that innervate distinct intrinsic laryngeal muscles to assess the role of CGRP in terms of the laryngeal functions [16].

10.3 Location of the Laryngeal Motoneurons in the NA

To identify the relative locations of the laryngeal motoneurons that innervate specific intrinsic laryngeal muscles, and to determine whether there exists the collateralization of laryngeal motoneurons to distinct types of muscles, we injected dual or triple retrograde fluorescent tracer [17] into the intrinsic laryngeal muscles in dogs [5, 6]. Cellular locations were assessed by immunohistochemistry in every consecutive section at the level of the NA, such that the number of neurons that projected to the specific intrinsic laryngeal muscles including the TA, PCA, LCA, Ary, and CT muscles can be counted. Two or three retrograde tracers were simultaneously injected into the different intrinsic laryngeal muscles, as shown in ◘ Table 10.1.

The number of neurons that innervated to the CT, PCA, or TA muscles was 100–300 and that innervated to the LCA or Ary muscles were 80–100, respectively. The rostrocaudal distribution of neurons that innervated to the TA, LCA, PCA, and Ary muscles were overlapped with each other, although the CT motoneurons were located to more rostral portion of the NA. The column of the neurons that innervated to the CT was extended from 1.5 mm caudal to the caudalmost part of the facial nucleus to the rostralmost part of the NA. These cells were mainly distributed at the level of the rostral part of the inferior olive and were interspersed within the area where the relatively large cells were observed. The neurons that projected to the PCA muscle were located at the level of between 1.0 mm caudal and 1.4 mm rostral to the obex. The laryngeal motoneurons that innervated to the TA muscle were located at the level of between 1.0 mm caudal to the obex and slightly caudal to the rostral margin of the PCA motoneurons pool. The cell column of the LCA motoneurons was located in between a level slightly caudal to the caudal end of the PCA motoneurons pool, which corresponds to the level of the caudal margin of the inferior olive, and a level slightly caudal to the rostral end of the TA motoneuron pool. The rostrocaudal extent of the motoneuron pool that projected to the Ary was approximately the same as that of the LCA motoneuron pool, whereas these motoneurons were located dorsally in the NA with reference to the other laryngeal motoneuron columns. We identified both the TA and PCA motoneurons in the serial sections of group A and B animals that a triple tracer injection to the intrinsic laryngeal muscles including the TA and PCA muscles was conducted. Regarding the dorsoventral coordinate, the motoneurons that projected to the TA muscle were located dorsally compared with those to the PCA muscle. The TA and LCA motoneurons were intermingled in the sections in group B at the rostrocaudal level corresponding to the overlapped region of these cell columns. Furthermore, we found some neurons in the NA in group B were double labeled by both DAPI and PI tracers, suggesting that these neurons have collateral axons projecting to the TA and LCA muscles (◘ Fig. 10.2). Otherwise, there was no other pattern of axonal collateralization among laryngeal motoneurons examined in this study. The above results were summarized in the diagram and the outline drawings by careful comparison of many photographic plates (◘ Fig. 10.3).

As reported in previous studies, the neurons that innervated to the CT were distributed more rostrally than those projecting the other type of intrinsic laryngeal muscles. This difference in localization of laryngeal motoneurons may be due to differences of branchial origin. The lateral branch of the superior laryngeal nerve and the CT are developed from fourth branchial arch structures, while intrinsic laryngeal muscles other than the CT and the recurrent laryngeal

◘ **Table 10.1** Injection muscles of three tracers in four groups [6]

Group	DAPI	PI	Pr
A	Thyroarytenoid	Cricothyroid	Posterior cricoarytenoid
B	Thyroarytenoid	Lateral cricoarytenoid	Posterior cricoarytenoid
C	Lateral cricoarytenoid	Arytenoid	
D	Arytenoid	Thyroarytenoid	

DAPI 4′,6-diamidino-2-phenylindol-2HCl, *PI* propidium iodide, *Pr* primuline

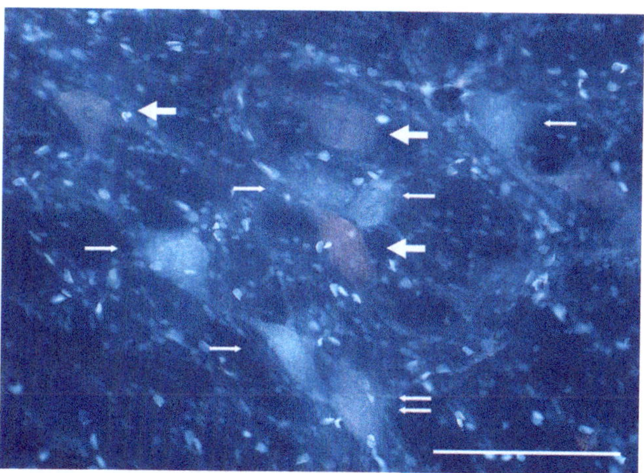

◘ **Fig. 10.2** Fluorescent micrograph of labeled cells in the nucleus ambiguus at the level just above the obex. Blue fluorescent cells labeled with DAPI injected into the thyroarytenoid muscle (←) were intermingled with the dim orange fluorescent cells labeled with PI injected into the lateral cricoarytenoid muscle (⬅) in the dorsal part of the nucleus, and the mixed blue and orange fluorescent cell doubly labeled with DAPI and PI (⇇) was also found. Bar = 100 μm [6]

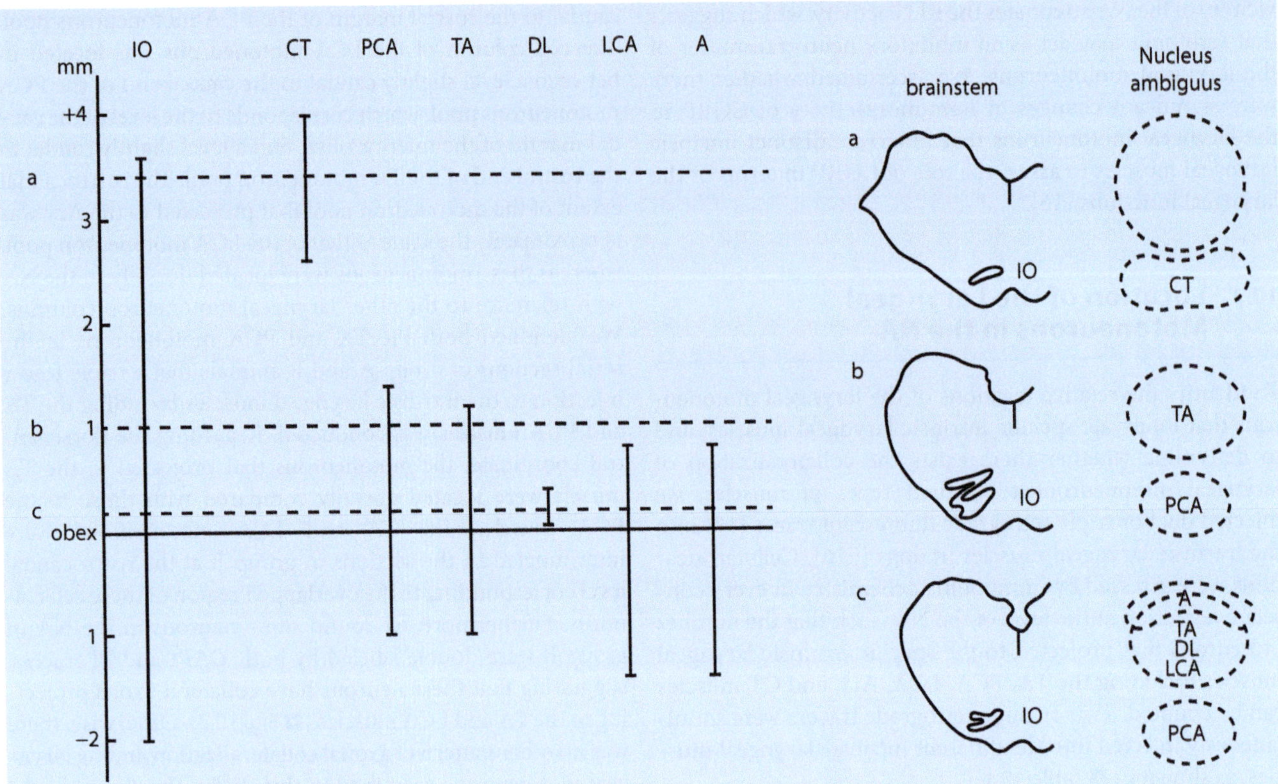

Fig. 10.3 A diagram showing the rostrocaudal extension of the labeled cell column in the nucleus ambiguus for each intrinsic laryngeal muscle of young dogs. Outline drawings show the transverse sections of the brainstem and the nucleus ambiguus taken through levels *a*, *b*, and *c* in the diagram. *A* arytenoid muscle, *CT* cricothyroid muscle, *DL* double-labeled cells from the thyroarytenoid and the lateral cricoarytenoid muscles, *IO* inferior olive, *LCA/PCA* lateral/posterior cricoarytenoid muscles, *TA* thyroarytenoid muscle [6]

nerve originate from the sixth branchial arch [18]. Yoshida et al. [3] have reported that the cell columns of the TA and PCA motoneurons were partly overlapped in the NA, whereas the area of TA motoneurons was located apart from that of PCA motoneurons in this study. This discrepancy may be attributed to the experimental settings or species differences. Although the suitable amount of the tracer injection into the individual intrinsic laryngeal muscle can avoid leakage of the tracer from the muscle itself, it seems difficult to delineate the precise boundary of these columns. With respect to the dorsoventral distribution of laryngeal motoneurons, the neurons that project to the laryngeal adductor muscles are located more rostrally than those that project to the laryngeal abductor muscle. In particular, the neurons whose axonal terminals were localized in the Ary muscle were located in the most dorsal aspect of the NA. One of the interesting findings of this study was the existence of double-labeled neurons whose axons projected to both the TA and LCA muscles [6], which may be also supported by the fact that the TA muscle is differentiated from the LCA muscle [18]. On the other hand, some investigators have noted that the TA muscle is subdivided by the caudomedial (referred to as vocalis muscle) and rostrolateral part [19–22]. Further studies will be necessary to reveal the histological features of the neurons that project to these distinct subdivisions.

10.4 CGRP Immunoreactivity of Neurons That Project to the Intrinsic Laryngeal Muscles

The vocal folds abduct during inspiration to intake air entering in the lung and then slightly adduct during early expiration to prevent lung collapse. This patterned activity can be generated by the brainstem neurons including laryngeal motoneurons. Since the physiological features of individual neuron might be influenced upon its intrinsic membrane property, it may be important to investigate what kind of neurotransmitter the neuron contains. The PCA motoneurons, which can only contribute to the inspiratory-related glottal opening, may be different from the other laryngeal motoneurons in immunoreactivity to the specific neurotransmitter. Indeed, approximately 70 % of the neurons in the NA that projected to the thyropharyngeal muscle, which contracts during swallowing, were immunopositive for CGRP, while only about 20 % of the neurons which innervated to the cricopharyngeal muscle, whose contraction is attenuated during the pharyngeal stage of swallowing to allow the bolus to enter the esophagus, were CGRP positive [23]. Therefore, to determine whether the laryngeal motoneurons contain CGRP as a neurotransmitter, and whether there is a relationship between the immunoreactivity of CGRP and the activity of each intrinsic laryngeal muscle, we performed immunohistochemistry of the laryngeal motoneurons retrogradely labeled by the cholera toxin B subunit-conjugated gold (CTBG) using the anti-CGRP antibody in dogs.

In this study, the somata of neurons that were immunopositive for CGRP showed no significant change in size and shape compared with that of the CGRP-negative laryngeal motoneurons. The labeled neurons (123–221 cells per animal) were identified in the rostral portion of the NA after CTBG injection into the CT muscle, most of which were immunopositive for CGRP (▣ Table 10.2). Meanwhile, the CTBG-labeled motoneurons that innervated to the TA or PCA muscle were located in the caudodorsal or caudoventral part of the NA, the number of which was 115–200 or 122–210 cells per animal, respectively. ▣ Figure 10.4 shows the morphology of neurons that were or were not labeled by CTBG injected into the PCA muscle, and those that were immunopositive or immunonegative for CGRP.

Although the localization of each motoneuron group was similar to that shown in the preliminary study, the number of labeled neurons was relatively large [6]. This may be due to the higher immunoreactivity of these neurons for CTBG than that for the fluorescent tracers utilized in our previous study.

▣ **Table 10.2 Ratio of motoneurons with CGRP immunoreactivity for each muscle [16]**

Muscle	Mean ± SD (%)
Cricothyroid muscle	93.0±0.6
Thyroarytenoid muscle	71.4±8.2
Posterior cricoarytenoid muscle	85.5±3.5

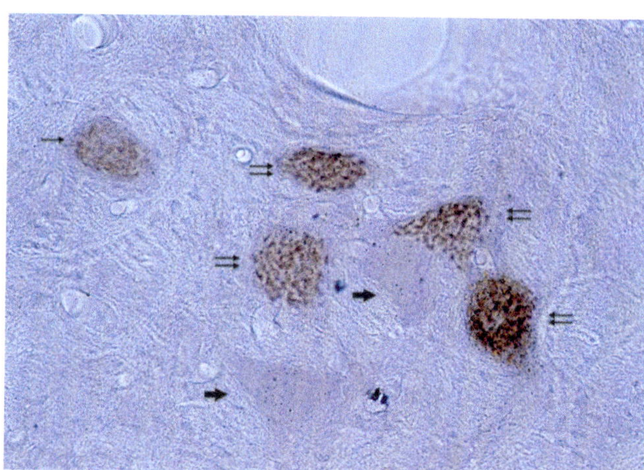

▣ **Fig. 10.4** Motoneurons in nucleus ambiguus after injection of cholera toxin B subunit-conjugated gold (CTBG) into posterior cricoarytenoid muscle. Motoneurons with only calcitonin gene-related peptide (CGRP) immunoreactivity were identified by presence of brown 3,3′ diaminobenzidine (DAB) products in their cytoplasm (→). Black grains resulting from silver intensification of CTBG were clearly distinguishable from DAB products obtained by immunohistochemistry. Thus, CTBG-labeled motoneurons without CGRP immunoreactivity (➡) and CTBG-labeled motoneurons with CGRP immunoreactivity (⇉) were clearly differentiated [16]

The positive ratio of motor neurons in the NA for CGRP immunohistochemistry is likely to depend on the neurons where the axon terminates. For example, 100, 98, 83, 92, 84, and 71 % of neurons whose axon projected to the heart, cervical esophagus, subdiaphragmatic esophagus, abdominal vagus nerve, recurrent laryngeal nerve, and stomach were immunopositive for CGRP in rats, respectively [24]. In addition, 70.3 and 21.2 % of CGRP-positive neurons that projected to the thyropharyngeal and cricopharyngeal muscles were identified in dogs, respectively [23]. Batten et al. [8] have proposed that the CGRP-positive neurons in the NA may be the motor neurons originated from the branchial arch and that the visceral efferent neurons in the NA may be immunonegative for CGRP. On the other hand, McWilliam et al. [9] have noted that only the motor neurons in the NA that project to the striated muscles were immunopositive for CGRP. Our results could not fully support these hypotheses.

The physiological role of CGRP in the motor nervous system is still unclear. Lee et al. [25] have proposed that CGRP functions as a co-transmitter with acetylcholine in vagal motoneurons including the CT motoneurons. Takami et al. [26] have also noted that CGRP is involved in the regulation of the striated muscle contraction. On the other hand, New and Mudge [27] suggested the possible role of CGRP in the neuromuscular junction that it may increase acetylcholine receptor synthesis by acting as a neurotrophic factor.

We thus conclude that, since the immunopositive rate for CGRP of the laryngeal motoneurons depended on which muscle the neurons project to, there may be a distinct predominance of the CGRP immunopositive neurons that innervate the different intrinsic laryngeal muscles regarding the neurotransmission or synaptic synthesis. In addition, CGRP might influence the regeneration of the recurrent laryngeal nerve when damaged, acting as a brain-derived trophic factor, although additional studies are required to address this issue.

References

1. Yoshida Y, Miyazaki T, Hirano M, Shin T, Totoki T, Kanaseki T. Location of motoneurons supplying the cricopharyngeal muscle in the cat studied by means of the horseradish peroxidase method. Neurosci Lett. 1980;18:1–4.
2. Yoshida Y, Miyazaki T, Hirano M, Shin T, Totoki T, Kanaseki T. Localization of efferent neurons innervating the pharyngeal constrictor muscles and the cervical esophagus muscle in the cat by means of the horseradish peroxidase method. Neurosci Lett. 1981;22:91–5.
3. Yoshida Y, Miyazaki T, Hirano M, Shin T, Kanaseki T. Arrangement of motoneurons innervating the intrinsic laryngeal muscles of cats as demonstrated by horseradish peroxidase. Acta Otolaryngol. 1982;94:329–34.
4. Yoshida Y, Tanaka Y, Saito T, Shimazaki T, Hirano M. Peripheral nervous system in the larynx. An anatomical study of the motor, sensory and autonomic nerve fibers. Folia Phoniatr (Basel). 1992;44:194–219.

5. Hisa Y, Matsui T, Sato F, Matsuura T, Fukui K, Tange A, Ibata Y. The localization of the motor neurons innervating the cricothyroid muscle in the adult dog by the fluorescent retrograde axonal labeling technique. Arch Otorhinolaryngol. 1982;234:33–6.
6. Hisa Y, Sato F, Fukui K, Ibata Y, Mizuokoshi O. Nucleus ambiguus motoneurons innervating the canine intrinsic laryngeal muscles by the fluorescent labeling technique. Exp Neurol. 1984;84:441–9.
7. Rosenfeld MG, Mermod JJ, Amara SG, Swanson LW, Sawchenko PE, Rivier J, Vale WW, Evans RM. Production of a novel neuropeptide encoded by the calcitonin gene via tissue-specific RNA processing. Nature. 1983;304:129–35.
8. Batten TF, Lo VK, Maqbool A, McWilliam PN. Distribution of calcitonin gene-related peptide-like immunoreactivity in the medulla oblongata of the cat, in relation to choline acetyltransferase-immunoreactive motoneurones and substance P-immunoreactive fibres. J Chem Neuroanat. 1989;2:163–76.
9. McWilliam PN, Maqbool A, Batten TF. Distribution of calcitonin gene-related peptide-like immunoreactivity in the nucleus ambiguus of the cat. J Comp Neurol. 1989;282:206–14.
10. Saji M, Miura M. Coexistence of glutamate and choline acetyltransferase in a major subpopulation of laryngeal motoneurons of the rat. Neurosci Lett. 1991;123:175–8.
11. Batten TF. Immunolocalization of putative neurotransmitters innervating autonomic regulating neurons of cat ventral medulla. Brain Res Bull. 1995;37:487–506.
12. McCrimmon DR, Feldman JL, Speck DF. Respiratory motoneuronal activity is altered by injections of picomoles of glutamate into cat brain stem. J Neurosci. 1986;6:2384–92.
13. King KA, Holtman Jr JR. Gamma-aminobutyric acid receptors at the ventral surface of the medulla inhibit respiratory motor outflow to the laryngeal musculature. Neuropharmacology. 1989;28:255–62.
14. Yajima Y, Hayashi Y. GABA(A) receptor-mediated inhibition in the nucleus ambiguus motoneuron. Neuroscience. 1997;79:1079–88.
15. King KA, Holtman Jr JR. Characterization of the effects of activation of ventral medullary serotonin receptor subtypes on cardiovascular activity and respiratory motor outflow to the diaphragm and larynx. J Pharmacol Exp Ther. 1990;252:665–74.
16. Hisa Y, Tadaki N, Koike S, Bamba H, Uno T. Calcitonin gene-related peptide-like immunoreactive motoneurons innervating the canine intrinsic laryngeal muscles. Ann Otol Rhinol Laryngol. 1998;107:1029–32.
17. Lawn AM. The nucleus ambiguus of the rabbit. J Comp Neurol. 1966;127:307–20.
18. Hast MH. Early development of the human laryngeal muscles. Ann Otol Rhinol Laryngol. 1972;81:524–30.
19. Sanders I, Han Y, Wang J, Biller H. Muscle spindles are concentrated in the superior vocalis subcompartment of the human thyroarytenoid muscle. J Voice. 1998;12:7–16.
20. Imamura R, Yoshida Y, Fukunaga H, Nakashima T, Hirano M. Thyroarytenoid muscle: functional subunits based on morphology and muscle fiber typing in cats. Ann Otol Rhinol Laryngol. 2001;110:158–67.
21. Mascarello F, Veggetti A. A comparative histochemical study of intrinsic laryngeal muscles of ungulates and carnivores. Basic Appl Histochem. 1979;23:103–25.
22. Yokoyama T, Nonaka S, Mori S. Histochemical properties of intrinsic laryngeal muscles in cats. J Auton Nerv Syst. 1995;56:50–60.
23. Hisa Y, Tadaki N, Uno T, Okamura H, Taguchi J, Ibata Y. Calcitonin gene-related peptide-like immunoreactive motoneurons innervating the canine inferior pharyngeal constrictor muscle. Acta Otolaryngol. 1994;114:560–4.
24. Stuesse SL, Sawchenko PE. Calcitonin generelated peptide and choline acetyltransferase : co-distribution in visceral subpopulations within the nucleus ambiguus of rat. Abstr Soc Neurosci. 1986;12:536.
25. Lee BH, Lynn RB, Lee HS, Miselis RR, Altschuler SM. Calcitonin gene-related peptide in nucleus ambiguus motoneurons in rat: viscerotopic organization. J Comp Neurol. 1992;320:531–43.
26. Takami K, Kawai Y, Uchida S, Tohyama M, Shiotani Y, Yoshida H, Emson PC, Girgis S, Hillyard CJ, MacIntyre I. Effect of calcitonin gene-related peptide on contraction of striated muscle in the mouse. Neurosci Lett. 1985;60:227–30.
27. New HV, Mudge AW. Calcitonin gene-related peptide regulates muscle acetylcholine receptor synthesis. Nature. 1986;323:809–11.
28. Miyazaki T. Somatotopical organization of the motoneurons in the nucleus ambiguus of cats – the horseradish peroxidase method. Jibi to Rinsyo. 1982;28:649–79.

Nucleus Tractus Solitarius

Shigeyuki Mukudai, Yoichiro Sugiyama, and Yasuo Hisa

S. Mukudai
Department of Otolaryngology-Head and Neck Surgery,
Kyoto Prefectural University of Medicine, Kawaramachi-Hirokoji,
Kamigyo-ku, Kyoto 602-8566, Japan
e-mail: s-muku@koto.kpu-m.ac.jp

Y. Sugiyama • Y. Hisa (✉)
Department of Otolaryngology-Head and Neck Surgery,
Kyoto Prefectural University of Medicine, Kawaramachi-Hirokoji,
Kamigyo-ku, Kyoto 602-8566, Japan
e-mail: yhisa@koto.kpu-m.ac.jp

Y. Hisa (ed.), *Neuroanatomy and Neurophysiology of the Larynx*, DOI 10.1007/978-4-431-55750-0_11

11.1 Nucleus Tractus Solitarius

The nucleus tractus solitarius (NTS), a longitudinal column of neurons in the dorsomedial medulla, is located along with the tractus solitarius that consists of myelinated afferent fibers of the general visceral afferent fibers of the facial, glossopharyngeal nerves and the general and special afferent fibers of the vagal nerve and that runs rostrocaudally from the caudal most level of the facial nucleus to the level of the caudal to the obex.

The NTS is subdivided into the following nine subnuclei based on their cellular morphology: commissural (paracommissural), gelatinous, medial, dorsal, dorsolateral, ventrolateral, ventral, intermediate, and interstitial subnuclei. Figure 11.1 shows the anatomical constitution of the subnuclei of the NTS in the transverse plane [1, 2]. In addition, the neurons of each subnucleus have a distinct morphological feature. For example, the commissural and paracommissural subnuclei consist of mainly small, bipolar-shaped neurons with clear, unstained cytoplasm, while the neurons, which are scattered in the gelatinous subnucleus, have the small- to medium-sized triangular- or rectangular-shaped somata. The interstitial subnucleus consists of the various-sized neurons, which are sparsely distributed in and around the tractus solitarius. As mentioned above, the general and special visceral afferent fibers of those cranial nerves convey inputs from various visceral organs. In particular, the neurons in the rostral NTS receive gustatory inputs from the facial and glossopharyngeal nerve, such that this area is called the gustatory nucleus.

On the other hand, the NTS also contributes to the regulation of the digestive and cardiovascular systems, from which the middle and caudal level of the NTS receive the sensory inputs, respectively. Moreover, the neurons in the NTS have the proprioceptive efferents to the various nuclei including the ventrolateral reticular formation, retrofacial nucleus, dorsal motor nucleus of the vagus, nucleus ambiguus, and nucleus retroambiguus, whereas the neurons in the NTS send their axons to the higher center including pons and thalamus. These interconnections could contribute to the following functions: (1) the respiratory, cardiovascular, and digestive system, (2) the pharyngeal and laryngeal reflexes, and (3) the hypothalamus-pituitary neuronal network regarding hormone secretion.

11.2 Neurotransmitters and Neuropeptides in the NTS

A variety of neurotransmitters and neuropeptides have been identified in the NTS neurons in which the heterogeneity of distribution of these substances has been demonstrated in the neuronal elements including somata, dendrites, and axons.

11.2.1 Catecholamine

A dense distribution of catecholaminergic fibers (i.e., noradrenaline, adrenaline, and dopamine) has been identified in the solitary tract [3–7]. On the other hand, it has been revealed that the caudal part of the NTS and the medial subnucleus contain rich population of the noradrenergic neurons, while the adrenergic neurons are mainly distributed in the medial and rostral portion of the NTS [3]. On the contrary, the adrenaline-containing fibers are predominantly located in the commissural subnucleus and are intermixed with the noradrenergic fibers broadly distributed in the NTS.

11.2.2 Substance P (SP)

Substance P-immunopositive neurons are broadly distributed in the NTS, in particular a dense population of these neurons can be observed in the rostromedial portion of the NTS [8]. Despite a sparse distribution of nerve fibers immunoreactive to substance P in the NTS, the diverse population of these fibers was observed in different species. For example, substance P-immunopositive fibers are densely distributed in the medial and commissural subnuclei of the NTS in rats [9], while these fibers are mainly located in the dorsolateral subnucleus of the NTS in felines [8].

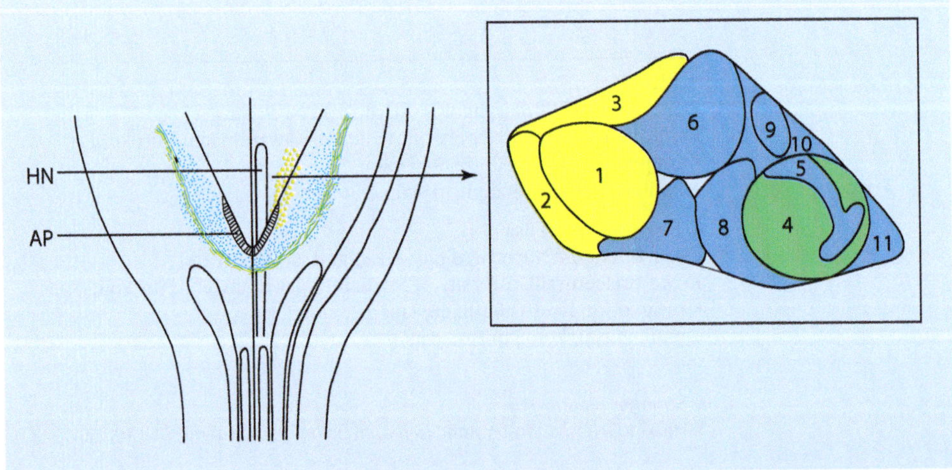

Fig. 11.1 The subnuclei of the nucleus tractus solitarius (right transverse plane) *HN* hypoglossal nucleus, *AP* area postrema, 1–3 dorsal motor nucleus of the vagus (*1* main part, *2* medial fringe, *3* dorsal fringe) *4* solitary tract, 5–11 nucleus tractus solitarius (*5* interstitial subnucleus, *6* medial subnucleus, *7* intermediate subnucleus, *8* ventral subnucleus, *9* dorsal subnucleus, *10* dorsolateral subnucleus, *11* ventrolateral subnucleus) (Revised from Ref. [2])

11.2.3 Calcitonin Gene-Related Peptide (CGRP)

Many neurons in the medial (e.g., commissural and medial subnuclei) and dorsal part of the NTS are immunoreactive to the CGRP. The CGRP-immunopositive fibers tend to be located in the caudolateral and medial part of the NTS, which are more sparsely distributed than SP-immunopositive fibers [10]. Many nerve fibers in the solitary tract are also immunopositive for the CGRP.

11.2.4 Enkephalin (ENK)

The ENK-immunopositive neurons are mainly located in the commissural and medial subnuclei of the NTS [11] and are scattered in the dorsolateral, ventrolateral, central, and ventral subnuclei [12]. The ENK-immunopositive fibers are broadly distributed in the NTS, especially in the commissural subnucleus; nevertheless, these fibers are less densely distributed than the SP-immunopositive fibers. In addition, these fibers may be associated with cardiovascular and respiratory regulation through the δ-opioid receptors [13].

11.3 Laryngeal Afferent Inputs to the NTS

Sweazey and Bradley [14] have reported that the superior laryngeal nerve (SLN), which conveys laryngeal sensory information, terminated in the central and ventral part of the rostral NTS, especially in the ventrolateral and interstitial subnuclei in sheep. In addition, projections of the laryngeal sensory afferents to the interstitial subnucleus have been reported in cats, rats, and hamsters [15–17]. Approximately 80 % of the laryngeal afferent fibers were projected to the specific region of the NTS (i.e., interstitial subnucleus), contributing to generating the pharyngeal and laryngeal reflexes to converge visceral signals to the NTS [18].

We examined the afferent projection of the inferior laryngeal nerve (ILN) in the NTS using the transganglionic tracer wheat germ agglutinin-horseradish peroxidase (WGA-HRP) in rats [16]. We transected the ILN at the level of the first tracheal ring and then immersed with WGA-HRP. Three days later, the brainstem was removed, and serial sections were stained with tetramethylbenzidine (TMB) reactions.

Sensory terminals of the afferent fibers in the ILN, containing WGA-HRP reaction product, were localized bilaterally in the area of the NTS and ipsilaterally in the tractus solitarius (◘ Fig. 11.2a). In the ipsilateral NTS, these terminals were found from 0.7 mm rostral to 0.8 mm caudal to the obex. The densest area of projection extended at the level of around 0.2 mm caudal to the obex, approximately corresponding to the longitudinal extension of the interstitial subnucleus (◘ Fig. 11.2b). A few labeled terminals were also found in the contralateral NTS, all of which were located in the interstitial subnucleus.

Our results indicated that the ILN, by which the intrinsic laryngeal muscles are innervated, also contains sensory afferent fibers and their projections to the NTS are heterogeneous and limited in localization.

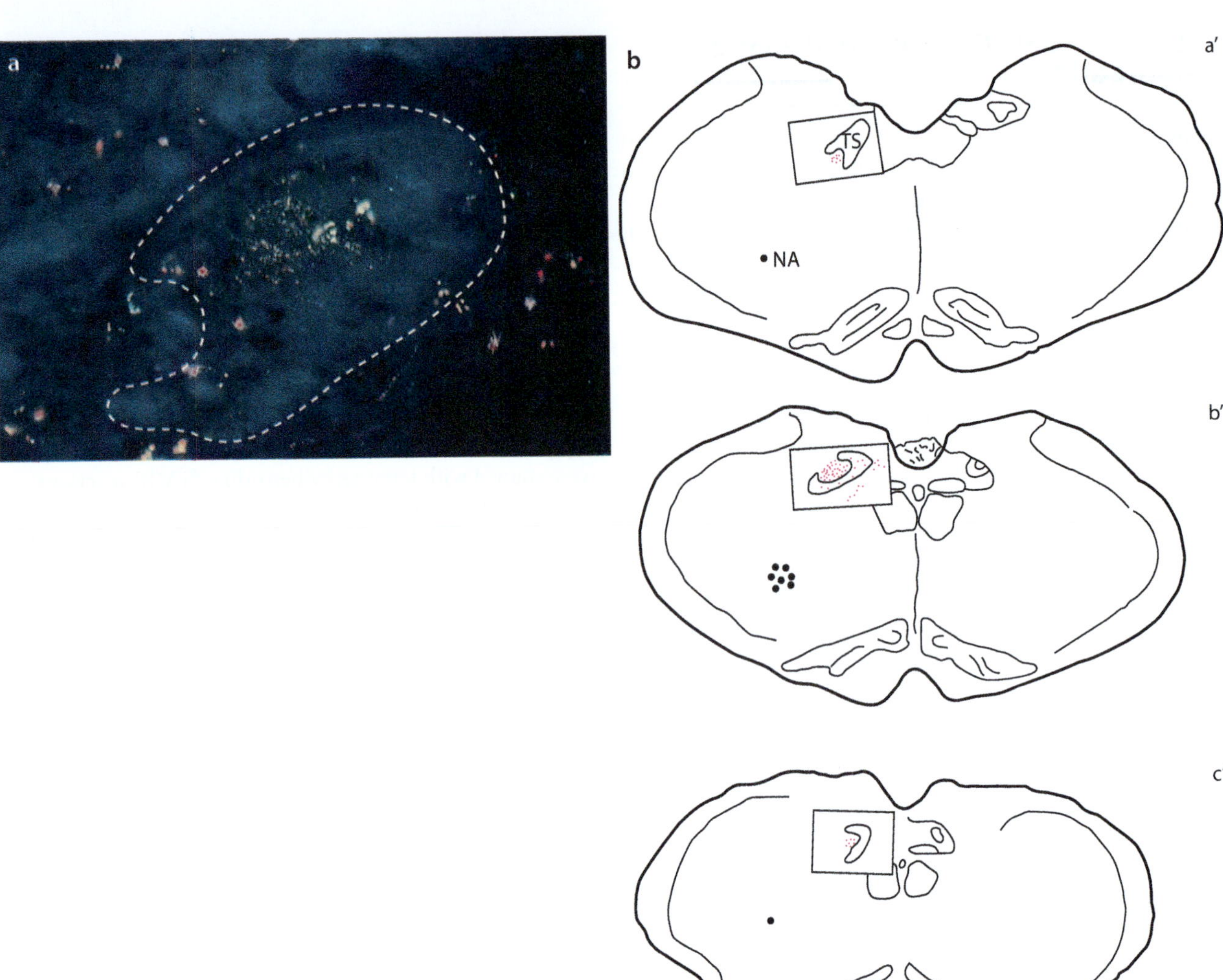

Fig. 11.2 (**a**) Dark-field photomicrograph shows the densest area of transganglionically labeled terminals within the right nucleus tractus solitarius after applying WGA-HRP to the ipsilateral inferior laryngeal nerve in the rat. Most of the terminals were localized in the interstitial subnucleus 0.2 mm caudal to the obex. Area inside *dotted line*: the nucleus tractus solitarius. (**b**) Camera lucida drawings of coronal sections through the rat medulla. The boxed area has been enlarged to show the area of labeled nucleus tractus solitarius. The most rostrally labeled terminals (0.7 mm rostral to the obex) were found in the area shown in a′, while the most caudally labeled terminals (0.8 mm caudal to obex) were found in the area shown in c′ However, the most prominent area of terminal label (0.2 mm caudal to obex) has been depicted in b′. *NA* nucleus ambiguus, *TS* tractus solitarius, *small dots* labeled terminals, *large dots* labeled motoneurons (Revised from Ref. [16])

References

1. Törk I, McRitchie DA, Rikard-Bell GC, Paxinos G. Autonomic regulatory centers in the medulla oblongata. In: Paxinos G, editor. The human nervous system. New York: Academic; 1990. p. 221–59.
2. Sano Y. Shinkei-Kagaku (Keitaigakuteki Kiso) Vol. 2. Spinal cord and brainstem. Kyoto: Kinpodo; 1999. p. 394–422.
3. Van der Gugten J, Palkovits M, Wijnen HL, Versteeg DH. Regional distribution of adrenaline in rat brain. Brain Res. 1976;107:171–5.
4. Versteeg DH, Van Der Gugten J, De Jong W, Palkovits M. Regional concentrations of noradrenaline and dopamine in rat brain. Brain Res. 1976;113:563–74.
5. Armstrong DM, Ross CA, Pickel VM, Joh TH, Reis DJ. Distribution of dopamine-, noradrenaline-, and adrenaline-containing cell bodies in the rat medulla oblongata: demonstrated by the immunocytochemical localization of catecholamine biosynthetic enzymes. J Comp Neurol. 1982;212:173–87.
6. Kalia M, Fuxe K. Rat medulla oblongata. I. Cytoarchitectonic considerations. J Comp Neurol. 1985;233:285–307.
7. Lambas-Senas L, Chamba G, Dennis T, Renaud B, Scatton B. Distribution of dopamine, noradrenaline and adrenaline in coronal sections of the rat lower brainstem. Brain Res. 1985;347:306–12.
8. Maley BE, Newton BW, Howes KA, Herman LM, Oloff CM, Smith KC, Elde RP. Immunohistochemical localization of substance P and enkephalin in the nucleus tractus solitarii of the rhesus monkey, Macaca mulatta. J Comp Neurol. 1987;260:483–90.
9. Yamazoe M, Shiosaka S, Shibasaki T, Ling N, Tateishi K, Hashimura E, Hamaoka T, Kimmel JR, Matsuo H, Tohyama M. Distribution of six neuropeptides in the nucleus tractus solitarii of the rat: an immunohistochemical analysis. Neuroscience. 1984;13:1243–66.
10. Berk ML, Smith SE, Karten HJ. Nucleus of the solitary tract and dorsal motor nucleus of the vagus nerve of the pigeon: localization of peptide and 5-hydroxytryptamine immunoreactive fibers. J Comp Neurol. 1993;338:521–48.

11. Kalia M, Fuxe K, Hökfelt T, Johansson O, Lang R, Ganten D, Cuello C, Terenius L. Distribution of neuropeptide immunoreactive nerve terminals within the subnuclei of the nucleus of the tractus solitarius of the rat. J Comp Neurol. 1984;222:409–44.
12. Lee HS, Basbaum AI. Immunoreactive pro-enkephalin and prodynorphin products are differentially distributed within the nucleus of the solitary tract of the rat. J Comp Neurol. 1984;230:614–9.
13. Hassen AH, Feuerstein GZ, Faden AI. Cardiovascular responses to opioid agonists injected into the nucleus of tractus solitarius of anesthetized cats. Life Sci. 1982;31:2193–6.
14. Sweazey RD, Bradley RM. Central connections of the lingual-tonsillar branch of the glossopharyngeal nerve and the superior laryngeal nerve in lamb. J Comp Neurol. 1986;245:471–82.
15. Nomura S, Mizuno N. Central distribution of afferent and efferent components of the glossopharyngeal nerve: an HRP study in the cat. Brain Res. 1982;236:1–13.
16. Hisa Y, Lyon MJ, Malmgren LT. Central projection of the sensory component of the rat recurrent laryngeal nerve. Neurosci Lett. 1985;55:185–90.
17. Hanamori T, Smith DV. Central projections of the hamster superior laryngeal nerve. Brain Res Bull. 1986;16:271–9.
18. Ootani S, Umezaki T, Shin T, Murata Y. Convergence of afferents from the SLN and GPN in cat medullary swallowing neurons. Brain Res Bull. 1995;37:397–404.

Dorsal Motor Nucleus of the Vagus

Shigeyuki Mukudai, Yoichiro Sugiyama, and Yasuo Hisa

S. Mukudai • Y. Sugiyama • Y. Hisa (✉)
Department of Otolaryngology-Head and Neck Surgery,
Kyoto Prefectural University of Medicine, Kawaramachi-Hirokoji,
Kamigyo-ku, Kyoto 602-8566, Japan
e-mail: s-muku@koto.kpu-m.ac.jp; yhisa@koto.kpu-m.ac.jp

Y. Hisa (ed.), *Neuroanatomy and Neurophysiology of the Larynx*, DOI 10.1007/978-4-431-55750-0_12

12.1 Introduction

The general visceral efferent fibers of the glossopharyngeal and vagus nerves originate from the dorsal motor nucleus of the vagus (DMNV), in which the parasympathetic preganglionic neurons are located.

12.2 Location

DMNV is constructed for a rostrocaudal extended cell column, which is located just ventral to the vagal triangle of the fourth ventricle and dorsal to the hypoglossal nucleus, and has a longer longitudinal extension than the hypoglossal nucleus. The axons of neurons in the DMNV run ventrolaterally, crossing through the spinal trigeminal tract and nucleus, and exit the medulla oblongata from between the inferior olivary complex and the inferior cerebellar peduncle.

12.3 Cellular Construction

In the DMNV, relatively small spindle cells and large cells in which coarse chromosomes and melanin pigment are interspersed coexist. DMNV contains the principal part of parasympathetic preganglionic neurons and peripheral dorsal and medial fringes. The principal part is located close to the nucleus tractus solitarius (NTS) and forms a cell column similar to the NTS in length. In the caudal part of the medulla, the DMNV is located in the ventral portion of the NTS and is incorporated into the subnuclei in the rostral NTS. The central part of the DMNV is divided into the dorsal region containing interspersed large neurons and compact ventral part of smaller neurons. In humans, there are 17,000–25,000 neurons in the bilateral DMNV [1].

12.4 Neurotransmitters

The principle neurotransmitter in the DMNV is acetylcholine (Ach) [2]. Several neurotransmitters have been found in the DMNV: catecholamine [3], 5-hydroxytryptamine [4], substance P [3–5], calcitonin gene-related peptide (CGRP) [4, 5], cholecystokinin [4], encephalin [4], neuropeptide Y [4], neurotensin [4], vasoactive intestinal polypeptide [4], galanin [6], somatostatin [4], oxytocin [7], thyrotropin-releasing hormone [6], cocaine- and amphetamine-regulated transcript peptide [8], and bombesin [9].

12.5 Innervation of the DMNV Neurons into the Larynx

Kalia and Mesulam [10] have reported that the retrogradely labeled neurons were found in the DMNV after injecting horseradish peroxidase (HRP) into the larynx in cats. Several investigators have also demonstrated the presence of the presympathetic neurons in the DMNV which innervated to the larynx in rats, dogs, and hamsters [11–15].

12.6 Cells of Origin of Parasympathetic Preganglionic Fibers in the Laryngeal Nervous System

We investigated the distribution and number of the DMNV neurons that projected to the parasympathetic ganglion of the larynx through the internal branch of the superior laryngeal nerve (SLNi) using the cholera toxin B subunit (CTB) as a retrograde neuronal tracer.

The CTB was applied around the distal end of the SLNi in the vicinity of the larynx in dogs. After a 4-day survival period, perfusion fixation of the brain stem was performed, and immunohistochemical analysis for CTB was conducted. The labeled cells were only observed in the ipsilateral side of the DMNV and the majority of these neurons generally had a bipolar, fusiform-shaped somata, while the multipolar cells were scattered (◘ Fig. 12.1) [16]. They were distributed 2.7–5.3 mm rostral to the obex, and the mean number of the labeled cells was 343 (◘ Fig. 12.2). Previous studies have shown that the somatotopic organization of neurons that projected to the parasympathetic innervation to the visceral organs seemed to be apparent [17–21]. The DMNV neurons whose axons travel through the SLN were located in approximately 2 mm rostrocaudal extent of the DMNV between the caudal pole of the facial nucleus and the rostral part of the hypoglossal nucleus, corresponding to the level of the motoneuron pool innervating to the cricothyroid muscle (CT) [22]. These results have indicated that the level of the longitudinal extension of parasympathetic preganglionic neurons whose axons run in the SLN is likely the same as that of the CT motoneurons and thus suggested that the neurons that control the larynx may be topographically organized. Furthermore, since the labeled neurons were observed only in the ipsilateral side, it seems unlikely that the efferent fibers of the DMNV neurons extend across the midline of the brain stem, as proposed in the previous studies [17–19, 23].

12.7 Parasympathetic Nervous System and Neurotransmitters of the Larynx

In order to determine what kinds of neurotransmitters are likely to be related to the presympathetic regulation of the larynx, we additionally investigated the neurotransmitters of parasympathetic preganglionic neurons which innervated to the larynx through the inferior laryngeal nerve (ILN) using a retrograde tracing technique with immunohistochemical staining for the markers for several neurotransmitters in dogs. The cholera toxin B subunit-conjugated gold (CTBG) was injected in the ILN at the level of the first tracheal ring.

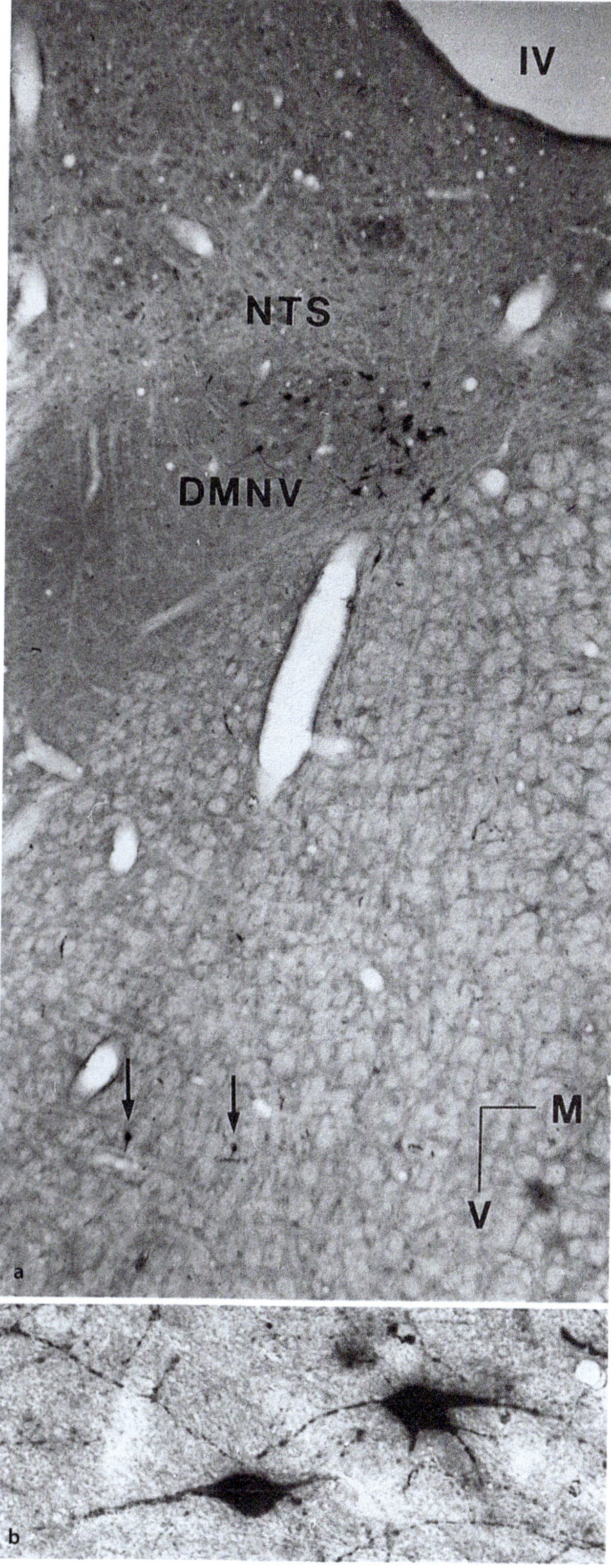

■ **Fig. 12.1** (**a**) The axial section at the level of 4.0 mm rostral to the obex. Labeled neurons were mainly found in the dorsomedial area in the ipsilateral dorsal motor nucleus of the vagus, and two labeled neurons (→) were found in the ipsilateral reticular formation. (**b**) Most of labeled neurons were bipolar, and some neurons were multipolar. *M* medial, *V* ventral, *DMNV* dorsal motor nucleus of the vagus, *NTS* nucleus tractus solitarius, *IV* fourth ventricle (Revised from Ref. [16])

Subsequent to the perfusion fixation, immunohistochemical analyses for choline acetyltransferase (ChAT) and CGRP as well as histochemical analysis for nicotinamide adenine dinucleotide phosphate diaphorase (NADPHd) were performed. The CTBG-labeled cells were located in the rostral part of the DMNV. Many neurons in the DMNV were ChAT-positive, so that most of the CTBG-labeled cells were immunopositive for the ChAT (■ Fig. 12.3a). Despite the lower density of CGRP immunopositive neurons than that of the ChAT-positive cells, some CTBG-labeled neurons were immunoreactive for the CGRP (■ Fig. 12.3b). The NADPHd-positive neurons could not be identified in the CTBG-labeled neurons (■ Fig. 12.3c).

Previous reports have demonstrated the existence of the parasympathetic ganglia which could control the laryngeal functions in and adjacent to the larynx [24, 25]. This study showed that the Ach- or CGRP-containing neurons in the DMNV projected to the parasympathetic ganglia in the larynx through the ILN and thus suggested that these neurotransmitters, which conveyed information to the parasympathetic neurons of the larynx, may play a pivotal role in the autonomic regulation of the larynx.

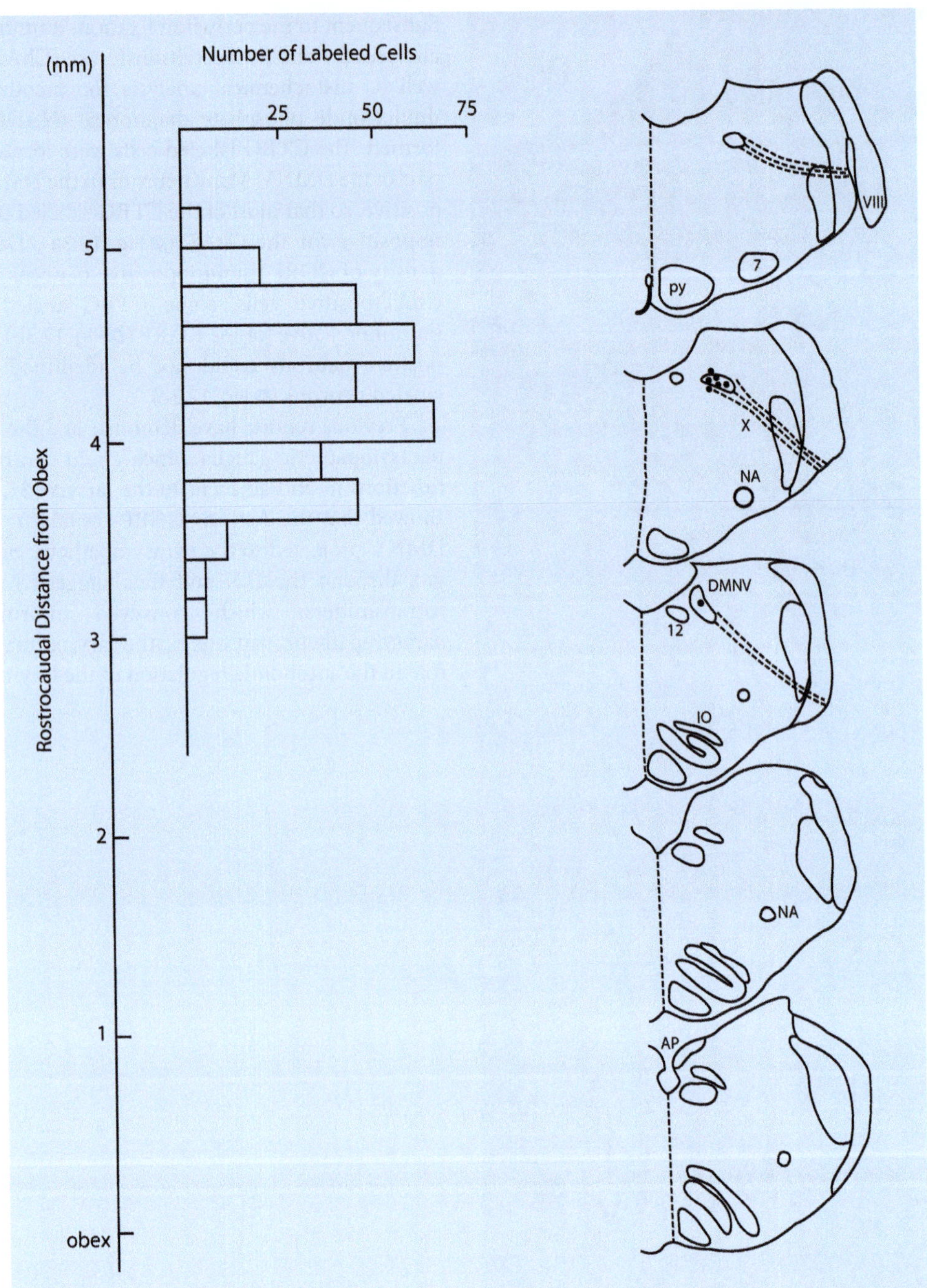

Fig. 12.2 The distribution of the label cells in the canine dorsal motor nucleus of the vagus (*left*) and the schematic diagram of the brainstem at each level (*right*). The labeled cells are found at 2.7–5.3 mm rostral to the obex. • labeled cell, *NA* nucleus ambiguus, *AP* area postrema, *DMNV* dorsal motor nucleus of the vagus, *IO* inferior olive, *PY* pyramidal tract, *7* facial nucleus, *VIII* vestibulocochlear nerve, *12* hypoglossal nucleus, *X* vagus nerve (Revised from Ref. [16])

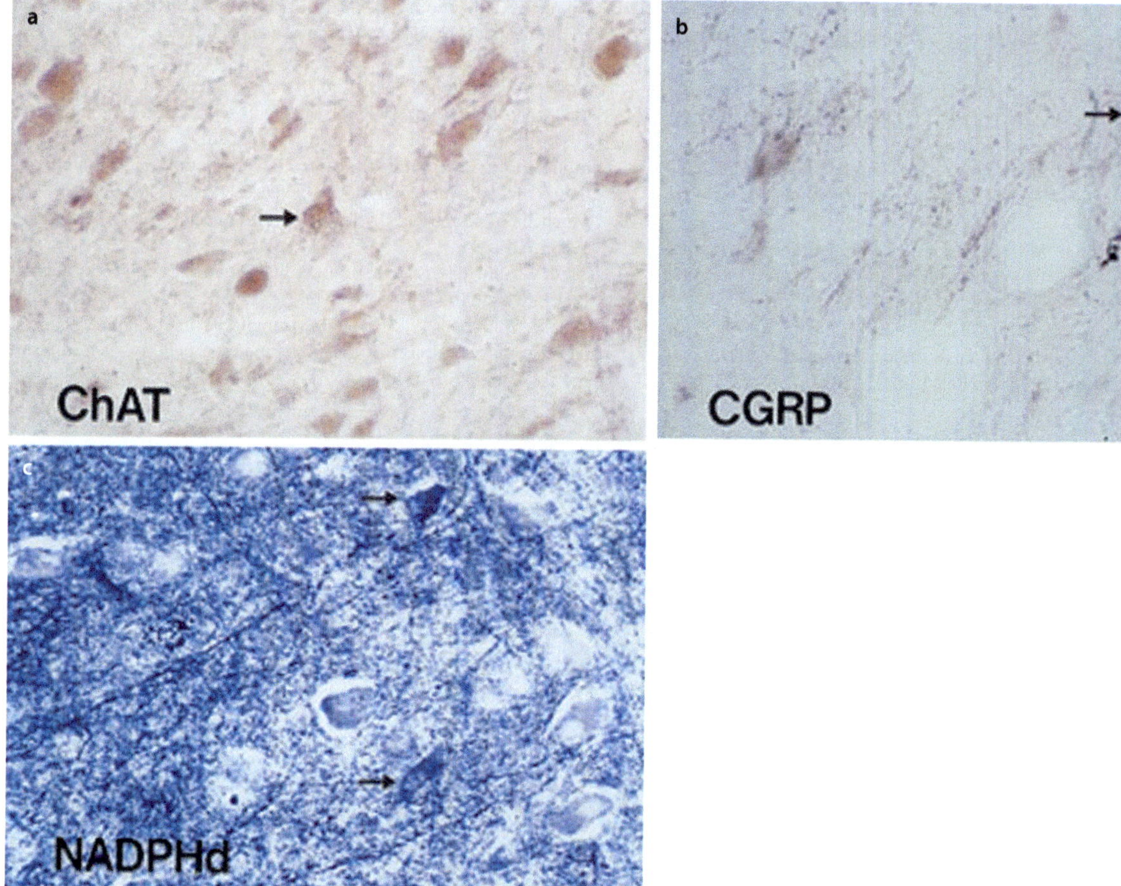

Fig. 12.3 The cholera toxin B subunit-conjugated gold (*CTBG*) was injected into the canine inferior laryngeal nerve, and double staining with choline acetyltransferase (*ChAT*), calcitonin gene-related peptide (*CGRP*), or nicotinamide adenine dinucleotide phosphate diaphorase (*NADPHd*) was performed. (**a**) CTBG-labeled neurons with ChAT immunoreactivity (→). Most of the DMNV cells were observed to be ChAT positive, and almost all CTBG-labeled cells were ChAT positive. (**b**) CTBG-labeled neurons with CGRP immunoreactivity (→). Few CGRP-positive cells were observed in the DMNV compared to ChAT-positive cells. (**c**) NADPHd-positive cells (→) and many NADPHd-positive fibers were observed in the DMNV, while NADPHd positivity was not observed in CTBG-labeled cells

References

1. Huang XF, Tork I, Paxinos G. Dorsal motor nucleus of the vagus nerve. A cyto- and chemoarchitectonic study in the human. J Comp Neurol. 1993;330:158–82.
2. Shute CC, Lewis PR. Cholinesterase-containing systems of the brain of the rat. Nature. 1963;199:1160–4.
3. Huang XF, Paxinos G, Halasz P, McRitchie D, Tork I. Substance P- and tyrosine hydroxylase-containing neurons in the human dorsal motor nucleus of the vagus nerve. J Comp Neurol. 1993;335:109–22.
4. Berk ML, Smith SE, Karten HJ. Nucleus of the solitary tract and dorsal motor nucleus of the vagus nerve of the pigeon: localization of peptide and 5-hydroxytryptamine immunoreactive fibers. J Comp Neurol. 1993;338:521–48.
5. Inagaki S, Kiyo S, Kubota Y, Girgis S, Hillyard CJ, MacIntyre I. Autoradiographic localization of calcitonin gene-related peptide binding sites in human and rat brains. Brain Res. 1986;374:287–98.
6. Fodor M, Pammer C, Gorcs T, Palkovits M. Neuropeptides in the human dorsal vagal complex: an immunohistochemical study. J Chem Neuroanat. 1994;7:141–57.
7. Dreifuss JJ, Raggenbass M, Charpak S, Dubois-Dauphin M, Tribollet E. A role of central oxytocin in autonomic functions: its action in the motor nucleus of the vagus nerve. Brain Res Bull. 1988;20:765–70.
8. Dun SL, Castellino SJ, Yang J, Chang JK, Dun NJ. Cocaine- and amphetamine-regulated transcript peptide-immunoreactivity in dorsal motor nucleus of the vagus neurons of immature rats. Brain Res Dev Brain Res. 2001;131:93–102.
9. Lynn RB, Hyde TM, Cooperman RR, Miselis RR. Distribution of bombesin-like immunoreactivity in the nucleus of the solitary tract and dorsal motor nucleus of the rat and human: colocalization with tyrosine hydroxylase. J Comp Neurol. 1996;369:552–70.
10. Kalia M, Mesulam MM. Brain stem projections of sensory and motor components of the vagus complex in the cat: II. Laryngeal, tracheobronchial, pulmonary, cardiac, and gastrointestinal branches. J Comp Neurol. 1980;193:467–508.
11. Hinrichsen CF, Ryan AT. Localization of laryngeal motoneurons in the rat: morphologic evidence for dual innervation? Exp Neurol. 1981;74:341–55.
12. Wallach JH, Rybicki KJ, Kaufman MP. Anatomical localization of the cells of origin of efferent fibers in the superior laryngeal and recurrent laryngeal nerves of dogs. Brain Res. 1983;261:307–11.
13. Hanamori T, Smith DV. Central projections of the hamster superior laryngeal nerve. Brain Res Bull. 1986;16:271–9.
14. Krammer EB, Streinzer W, Millesi W, Ellbock E, Zrunek M. Central localization of motor components in the internal branch of the superior laryngeal nerve: a horseradish peroxidase study in the rat. Laryngol Rhinol Otol (Stuttg). 1986;65:617–20.

15. Basterra J, Chumbley CC, Dilly PN. The superior laryngeal nerve: its projection to the dorsal motor nucleus of the vagus in the guinea pig. Laryngoscope. 1988;98:89–92.
16. Uno T. Autonomic neurons sending fibers into the canine laryngeal nerves using a retrograde tracer technique with cholera toxin. Nippon Jibiinkoka Gakkai Kaiho. 1993;96:66–76.
17. Hiura T. Salivatory neurons innervate the submandibular and sublingual glands in the rat: horseradish peroxidase study. Brain Res. 1977;137:145–9.
18. Contreras RJ, Gomez MM, Norgren R. Central origins of cranial nerve parasympathetic neurons in the rat. J Comp Neurol. 1980;190:373–94.
19. Nomura S, Mizuno N. Central distribution of afferent and efferent components of the chorda tympani in the cat as revealed by the horseradish peroxidase method. Brain Res. 1981;214:229–37.
20. Nomura S, Mizuno N. Central distribution of afferent and efferent components of the glossopharyngeal nerve: an HRP study in the cat. Brain Res. 1982;236:1–13.
21. Nomura S, Mizuno N. Central distribution of efferent components in the greater petrosal nerve of the cat. Neurosci Lett. 1983;39:11–4.
22. Hisa Y, Sato F, Fukui K, Ibata Y, Mizuokoshi O. Nucleus ambiguus motoneurons innervating the canine intrinsic laryngeal muscles by the fluorescent labeling technique. Exp Neurol. 1984;84:441–9.
23. Mitchell GA, Warwick R. The dorsal vagal nucleus. Acta Anat (Basel). 1955;25:371–95.
24. Tsuda K, Shin T, Masuko S. Immunohistochemical study of intralaryngeal ganglia in the cat. Otolaryngol Head Neck Surg. 1992;106:42–6.
25. Tanaka Y, Yoshida Y, Hirano M. Ganglionic neurons in vagal and laryngeal nerves projecting to larynx, and their peptidergic features in the cat. Acta Otolaryngol Suppl. 1993;506:61–6.

Central Projections to the Nucleus Ambiguus

Shigeyuki Mukudai, Yoichiro Sugiyama, and Yasuo Hisa

S. Mukudai • Y. Sugiyama • Y. Hisa (✉)
Department of Otolaryngology-Head and Neck Surgery, Kyoto Prefectural University of Medicine,
Kawaramachi-Hirokoji, Kamigyo-ku, Kyoto 602-8566, Japan
e-mail: s-muku@koto.kpu-m.ac.jp; yhisa@koto.kpu-m.ac.jp

Y. Hisa (ed.), *Neuroanatomy and Neurophysiology of the Larynx*, DOI 10.1007/978-4-431-55750-0_13

13.1 Introduction

The larynx plays an essential role in respiration, phonation, and airway protective reflexes including deglutition, coughing, and sneezing. In order to achieve these functionally distinct movements, well-coordinated movements of the muscles of the oral cavity, pharynx, chest and abdomen are necessary. It thus seems likely that there appears to be the complex neuronal circuitry that regulates the laryngeal functions, such that there must exist various descending projections to the laryngeal motoneurons. The nucleus ambiguus (NA) is located in the ventral medulla, constructing a longitudinal cell column. In particular, the motoneurons that project to the intrinsic laryngeal musculatures are mainly distributed in its caudal portion where the neurons are sparsely distributed (referred to as the loose formation). Due to these anatomical characteristics, it is extremely difficult to selectively inject neural tracers into the laryngeal motoneurons. This technical problem may interfere with the analysis of the morphological features of neurons which have mono- or polysynaptic inputs to the laryngeal motoneurons. Consequently, many investigators have determined their projections using not only the anatomical tracing techniques but also the physiological or pharmacological procedures. Otherwise, for a dense cluster of neurons in the NA (e.g., the esophageal motoneurons), the tracer microinjection technique has been utilized in the previous studies [1–3].

13

13.2 Projection to the Laryngeal Motoneurons

The monosynaptic neuronal pathway has been identified from the nucleus tractus solitarius (NTS) to the NA in which the motoneurons innervating the intrinsic laryngeal muscles were included [4].

It has also been demonstrated that the direct projections from the neurons in the nucleus retroambiguus (NRA) to the laryngeal motoneurons in the NA [5]. In addition, Boers et al. have proposed that the premotor neurons in the NRA that control the cricothyroid (CT) muscle activity are generally excitatory [6].

Besides, Arita et al. have reported the activation of the posterior cricoarytenoid (PCA) muscle induced by delivering the electrical stimulation to the nucleus raphe pallidus that was eliminated after application of the antagonists of the serotonergic receptor 5-HT2 and thus suggested that the PCA motoneurons possibly receive serotonergic inputs from neurons in the nucleus raphe pallidus [7]. They have also addressed the issue that direct microinjection of the retrograde tracer cholera toxin B subunit (CTB) into the NA can be expanded in the vicinity of the NA, resulting in difficulty in determining the exact projections to the NA. Meanwhile, they have indicated that the electrical stimulation of the amygdala and the lateral aspect of the hypothalamus evoked activity of the CT muscle, which is predominantly activated in the inspiratory phase of eupnea, in the early and late expiratory phase, respectively. The response patterns of the CT muscle activity to these stimulations may indicate that there might be the interneurons that relay signals from the limbic system including the amygdala and hypothalamus to the CT motoneurons, as they suggested [8].

13.3 Transsynaptic Tracing Studies

Since retrograde transsynaptic tracing technique with pseudovirus infection is able to reveal the polysynaptic projections to the specific target, several investigators have utilized the pseudorabies virus (PRV) to identify the location of neurons that provide inputs to the motoneurons in the NA [9–12]. In terms of the multisynaptic circuitry innervating the intrinsic laryngeal musculatures, no previous studies have systematically identified the networks of neuroanatomical structures, although the hierarchy of polysynaptic pathway to the specific laryngeal muscles such as the CT [10], PCA [9, 12], and the thyroarytenoid (TA) muscle [13] has been studied.

For example, Barret et al. [10] have reported that subsequent to injection of the PRV into the CT muscle, the neurons infected were labeled initially in the rostral part of the NA followed by the interstitial and central subnuclei of the NTS in a time-dependent manner. Fay et al. [9] have demonstrated that labeled neurons were distributed in the NA, the ventrolateral and lateral subnuclei of the NTS, the paratrigeminal nucleus, the reticular nucleus, and the raphe nucleus, after PRV injection into the PCA muscle. Since the level of declaration of the hierarchical network pathway could depend upon the persistence period of the viral infection, the other areas identified by Waldbaum et al. [12] such as the NRA, the periaqueductal gray (PAG), the hippocampus, the cortical motor area, the hypothalamus, and the amygdala seem to be more upstream regions providing inputs to the PCA motoneurons. Van Daele et al. [13] have reported that PRV followed three interconnected systems originating in the forebrain: a bilateral system including the ventral anterior cingulate cortex, PAG, and ventral respiratory group; an ipsilateral system involving the parietal insular cortex, central nucleus of the amygdala, and parvicellular reticular formation; and a minor contralateral system originating in the motor cortex, after PRV injection into the TA muscle.

13.4 Central Projections to the NA

No systematic study has investigated the trajectories of each intralaryngeal muscle-control neuron using a viral tracer. Therefore, we studied the projections on the NA using herpes simplex virus 2 (HSV-2) as a tracer [14]. We focused on the CT and TA/LCA muscles in male ICR mice. Due to difficulty in clearly distinguishing between TA and LCA muscles for injection, one group comprised of both the muscles. Under deep anesthesia, the neck was excised open to expose the larynx, and HSV-2 was injected into one side of the CT or TA/LCA muscles. In order to block virus transportation via sen-

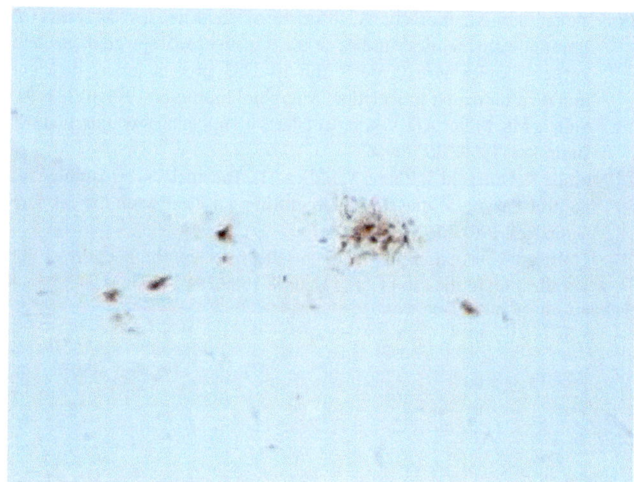

Fig. 13.1 HSV-2-labeled cells in the area postrema. Labeled cells were observed bilaterally. The number of labeled cells was dominant on the injected (*right*) side

sory and sympathetic nerves, the cervical sympathetic trunk was removed in all mice, and the superior laryngeal nerve was cut in the TA/LCA group. The whole brain was removed to prepare frozen serial sections. We performed an immunohistochemical analysis on the sections with an anti-HSV-2 antibody.

Labeled cells were observed in the following nuclei:

CT group

(a) NA: Labeled cells were observed only in the ipsilateral NA.
(b) NRA: A few labeled cells were observed bilaterally.
(c) NTS: Labeled cells were observed bilaterally. They were observed in interpolar, central, and lateral subnuclei, in addition to interstitial subnucleus where sensory fibers project from the larynx.
(d) Reticular nucleus: Relatively small cells were labeled around the NA. The labeled cells were observed in the ventral as well as dorsal side of the NA and observed not only on the injected side but also on the non-injected side.
(e) Paratrigeminal nucleus: Labeled cells were observed bilaterally.
(f) Area postrema: Labeled cells were observed bilaterally. The number of labeled cells was dominant on the injected side (Fig. 13.1).
(g) Vestibular nucleus: Labeled cells were observed bilaterally. They were observed in the medial as well as lateral vestibular nucleus. The number of labeled cells was dominant on the injected side.
(h) Above midbrain: No labeled cells were observed.

TA/LCA group

(a) NA: Labeled cells were observed only in the ipsilateral NA.
(b) NRA: A few labeled cells were observed bilaterally.
(c) NTS: Labeled cells were observed bilaterally. They were observed in lateral and interpolar subnuclei, in addition to interstitial subnucleus where sensory fibers project from the larynx.
(d) Reticular nucleus: Relatively small cells were labeled around the NA.
(e) Paratrigeminal nucleus: Labeled cells were observed bilaterally.
(f) Area postrema: Labeled cells were observed bilaterally.
(g) Vestibular nucleus: Few Labeled cells were observed bilaterally, compared to the CT group. They were observed in the medial as well as lateral vestibular nucleus. The number of labeled cells was dominant on the injected side.
(h) Cochlear nucleus: Labeled cells were observed bilaterally. They were observed in the dorsal as well as ventral cochlear nucleus.
(i) Above midbrain: No labeled cells were observed.

These results do not contradict previous reports using physiological techniques [15, 16]. We have provided novel evidence for communication between the cochlear nucleus and the NA. Further studies are required to ascertain whether the labeled cells observed in this study were only primary and secondary neurons or tertiary and higher-level neurons as well. However, although the PAG, which is highly associated with phonation function, is thought to contain tertiary neurons belonging to the PAG-NRA-NA pathway [17], in this study, labeled cells were not observed in the PAG, which suggests that the labeled cells were primary and secondary neurons. We aim to further clarify the neural mechanism controlling laryngeal function by varying factors such as tracers and lifetime of animals.

References

1. Bieger D. Muscarinic activation of rhombencephalic neurones controlling oesophageal peristalsis in the rat. Neuropharmacology. 1984;23:1451–64.
2. Ross CA, Ruggiero DA, Reis DJ. Projections from the nucleus tractus solitarii to the rostral ventrolateral medulla. J Comp Neurol. 1985;242:511–34.
3. Cunningham Jr ET, Sawchenko PE. A circumscribed projection from the nucleus of the solitary tract to the nucleus ambiguus in the rat: anatomical evidence for somatostatin-28-immunoreactive interneurons subserving reflex control of esophageal motility. J Neurosci. 1989;9:1668–82.
4. Hayakawa T, Takanaga A, Maeda S, Ito H, Seki M. Monosynaptic inputs from the nucleus tractus solitarii to the laryngeal motoneurons in the nucleus ambiguus of the rat. Anat Embryol (Berl). 2000;202:411–20.
5. Zhang SP, Bandler R, Davis PJ. Brain stem integration of vocalization: role of the nucleus retroambigualis. J Neurophysiol. 1995;74:2500–12.
6. Boers J, Klop EM, Hulshoff AC, de Weerd H, Holstege G. Direct projections from the nucleus retroambiguus to cricothyroid motoneurons in the cat. Neurosci Lett. 2002;319:5–8.
7. Arita H, Ichikawa K, Sakamoto M. Serotonergic cells in nucleus raphe pallidus provide tonic drive to posterior cricoarytenoid motoneurons via 5-hydroxytryptamine2 receptors in cats. Neurosci Lett. 1995;197:113–6.
8. Arita H, Kita I, Sakamoto M. Two distinct descending inputs to the cricothyroid motoneuron in the medulla originating from the amygdala and the lateral hypothalamic area. Adv Exp Med Biol. 1995;393:53–8.
9. Fay R, Gilbert KA, Lydic R. Pontomedullary neurons transsynaptically labeled by laryngeal pseudorabies virus. Neuroreport. 1993;5:141–4.

10. Barrett RT, Bao X, Miselis RR, Altschuler SM. Brain stem localization of rodent esophageal premotor neurons revealed by transneuronal passage of pseudorabies virus. Gastroenterology. 1994;107:728–37.
11. Bao X, Wiedner EB, Altschuler SM. Transsynaptic localization of pharyngeal premotor neurons in rat. Brain Res. 1995;696:246–9.
12. Waldbaum S, Hadziefendic S, Erokwu B, Zaidi SI, Haxhiu MA. CNS innervation of posterior cricoarytenoid muscles: a transneuronal labeling study. Respir Physiol. 2001;126:113–25.
13. Van Daele DJ, Cassell MD. Multiple forebrain systems converge on motor neurons innervating the thyroarytenoid muscle. Neuroscience. 2009;162:501–24.
14. Yamamoto M, Kurachi R, Morishima T, Kito J, Nishiyama Y. Immunohistochemical studies on the transneuronal spread of virulent herpes simplex virus type 2 and its US3 protein kinase-deficient mutant after ocular inoculation. Microbiol Immunol. 1996;40:289–94.
15. Siniaia MS, Miller AD. Vestibular effects on upper airway musculature. Brain Res. 1996;736:160–4.
16. Shiba K, Umezaki T, Zheng Y, Miller AD. The nucleus retroambigualis controls laryngeal muscle activity during vocalization in the cat. Exp Brain Res. 1997;115:513–9.
17. Holstege G. Anatomical study of the final common pathway for vocalization in the cat. J Comp Neurol. 1989;284:242–52.

Neurophysiological Study of the Brain Stem

Central Pattern Generators

Yoichiro Sugiyama, Shinji Fuse, and Yasuo Hisa

Y. Sugiyama • S. Fuse • Y. Hisa (✉)
Department of Otolaryngology-Head and Neck Surgery,
Kyoto Prefectural University of Medicine, Kawaramachi-Hirokoji, Kamigyo-ku,
Kyoto 602-8566, Japan
e-mail: yoichiro@koto.kpu-m.ac.jp; yhisa@koto.kpu-m.ac.jp

Y. Hisa (ed.), *Neuroanatomy and Neurophysiology of the Larynx*, DOI 10.1007/978-4-431-55750-0_14

14.1 Brainstem Mechanisms Underlying Laryngeal Movements

Brainstem functions as not only a relay station of descending inputs from higher center but also pattern-generating system which can provide suitable reactions to protect ourselves from risks and to maintain homeostasis.

Laryngeal movements, such as breathing, phonation, and airway protective reflexes including swallowing and coughing, can be generated and controlled by the specific neuronal networks in the brainstem, which can be influenced by descending signals from higher center. These specific neuronal networks are called as "the central pattern generators (CPGs)."

The brainstem, where the CPG networks involved in the laryngeal movements exist, is anatomically classified as three subdivisions: (1) medulla oblongata, (2) pons, and (3) midbrain. Each area includes the specific neuronal groups that could participate in the generation of these behaviors.

For example, the cranial motoneurons including the laryngeal, pharyngeal, esophageal, and hypoglossal motoneurons, which can contribute to the generation of the laryngeal motor activities including breathing, vocalization, and airway protective reflexes, are located in the medulla. In particular, the laryngeal motoneurons are located in the loose formation of the nucleus ambiguus (NA), whereas the pharyngeal motoneurons are distributed in the semicompact formation of the NA. On the other hand, the neurons that project to the lumbar spinal cord or the NA are located in the nucleus retroambiguus (NRA), presumably acting as the premotor neurons of the abdominal or laryngeal motoneurons, respectively [1, 2]. The afferent of the upper airway and alimentary tract terminates in the nucleus tractus solitarius (NTS) and spinal trigeminal nucleus [3]. The midbrain periaqueductal gray (PAG) contributes significantly to the generation of vocalization [4].

The neuronal networks of the central respiratory pattern generator mainly exist in the medulla and pons. The respiratory neurons located in the ventrolateral NTS and adjacent reticular formation are known as the dorsal respiratory group (DRG) [5, 6]. On the other hand, the respiratory neurons in the ventrolateral medulla and pons constitute a longitudinal column extending from the facial nucleus to the rostralmost part of the cervical spinal cord in the lateral tegmental field. This column is subdivided into the following regions: (1) the retrotrapezoid/parafacial respiratory group (RTN/pFRG) anatomically corresponding to the ventrolateral to the facial nucleus, (2) the Bötzinger complex (BötC) located in the retrofacial nucleus and surrounding reticular formation (RF), (3) the pre-Bötzinger complex (pre-BötC) located just caudal to the retrofacial nucleus, (4) the rostral ventral respiratory group (rVRG) localized at the level of the NA, and (5) the caudal ventral respiratory group (cVRG) corresponding to the level of the NRA [7]. In addition, the pontine respiratory group (PRG) is located in the dorsolateral pons (▫ Fig. 14.1) [8].

The physiological and anatomical organization of the CPGs regarding various laryngeal movements is still not fully understood. In the following sections, we addressed the issue how brainstem neuronal networks contribute to the generation of the laryngeal movements including respiration, vocalization, swallowing, and coughing.

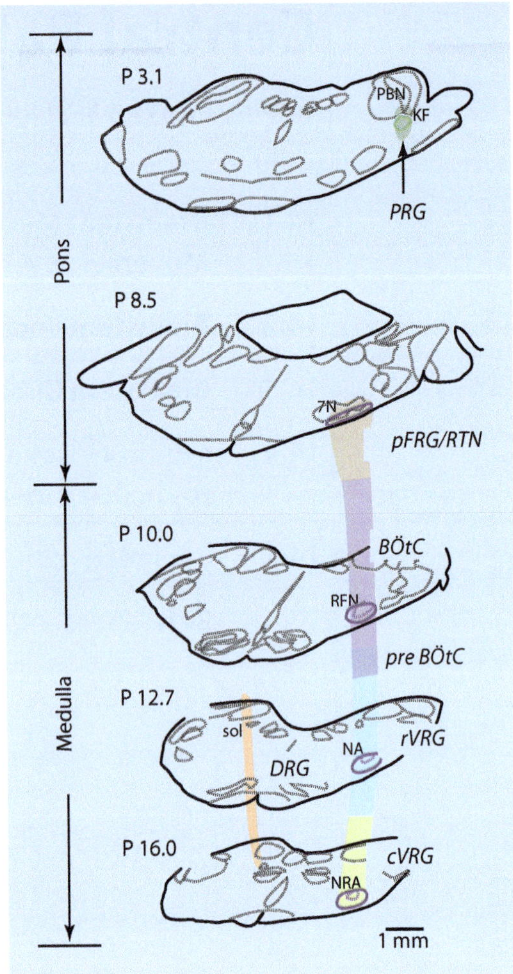

▫ **Fig. 14.1** Schematic drawing of the respiratory centers in the brainstem. *Colors* indicated in the transverse sections with reference to Berman's atlas [56] represent rostrocaudal extent of the respiratory groups including pontine respiratory group (PRG), parafacial respiratory group/retrotrapezoid nucleus (pFRG/RTN), the Bötzinger complex (BötC), the pre-Bötzinger complex (pre-BötC), the rostral ventral respiratory group (rVRG), the caudal ventral respiratory group (cVRG), and the dorsal respiratory group (DRG). In addition, numbers at the top of each transverse section indicate the level posterior (P) to stereotaxic zero. *KF* Kölliker-Fuse nucleus, *NA* nucleus ambiguus, *NRA* nucleus retroambiguus, *RFN* retrofacial nucleus, *sol* solitary tract, *7N* facial nucleus

14.2 Brainstem Vocalization Area

Human vocalization is produced by forced expiration accompanied by glottal closure being enhanced by resonance effect of nasal and pharyngeal cavity. In the animal model, vocalization is also consisted of the patterned movements of the vocal fold adduction and tension with abdominal constriction subsequent to inhalation. The PAG plays an important role in the generation of this patterned motion, since mutism can be attributed to lesion of the PAG [9–12]. As such, many investigators have focused on the physiological and anatomical role in the PAG in terms of how vocal movements can be generated. Electrical or chemical stimulation of the PAG evokes vocalization in monkeys and felines [13–15]. Tract-tracing studies have also revealed the direct connections from the PAG to the NRA, which can act as the final common pathway of PAG-induced vocalization [16]. Furthermore, as reported by Shiba et al. [17], dysfunction of the NRA abolished vocalization evoked by stimulation of the PAG, suggesting that this common pathway could be critical to produce PAG-induced vocalization.

We have also studied the brainstem vocal area and established fictive vocalization model in guinea pigs, in order to compare the mechanisms of brainstem vocalization with in other animals and to elucidate whether guinea pigs can be substitute for those animals to investigate brainstem mechanism underlying vocalization [18].

We first employed electrical stimulation from the midbrain to the lower brainstem systematically, such that we identified auditory vocalization during stimulation at the specific sites (◘ Fig. 14.2). Although guinea pigs can produce four typical vocalization calls, purr, chatter, chirp, and whistle, PAG-induced vocalization can represent two types of call: purr and whistle [19–22]. In this study, the call site stimulation could only produce the low whistle sound, probably because of the experimental setting. These call sites were distributed continuously from the lateral PAG to the ventromedial medulla at the level of the NA via the lateral part of the pontine reticular formation (◘ Fig. 14.2). Although this "PAG-medulla call area" did not continue to the caudal medulla, this area corresponded to the vocal pathway in other animals, which suggests that the vocal animals could possess the similar neuronal pathway involved in vocalization. On the other hand, our study showed chemical stimulation could evoke vocalization not only to the PAG but also to the pontine reticular formation and parabrachial region. There appear to be slight differences between the call sites that evoked by chemical stimuli in guinea pigs and those in other animals. For example, on the basis of our results, application of the excitatory amino acid did not evoke vocal reaction in guinea pigs in the area including the midbrain tegmentum and the pontine paralemniscus area where chemical stimuli can evoke vocalization in monkeys and bats, respectively [14, 23]. These differences may be attributed to the discrete extension of vocal center or sparsely distributed vocal-related cells in guinea pigs. As described above, the NRA is thought to be a critical area underlying vocalization especially adductor activity during the expiratory phase of vocalization [17]. Our data support this hypothesis, since we found that chemical stimuli in the vicinity of the NRA exhibited rhythmic activity of the vocal adductor muscle (◘ Fig. 14.3).

Again, in order to get to the bottom of the vocal CPG, it is necessary to study the cellular and network properties of the vocal-related neurons in the brainstem. Therefore, we then established fictive vocalization model using paralyzed guinea pigs, which represented the specific features of bursting activity of the superior laryngeal nerve (SLN), the abdominal nerve (ABD), followed by the phrenic nerve (PHR) activation (◘ Fig. 14.4). Consequently, we have established the animal model for investigating brainstem vocal mechanism in guinea pigs.

14.3 Brainstem Circuitry Involved in Swallowing

Swallowing is generated by spatially and temporally coordinated muscle contraction of oral cavity, pharynx, larynx, and esophagus, resulting in successful transition of the bolus without aspiration. These stereotyped movements are controlled by a consequence of the network activity of the swallowing CPG (Sw-CPG). The neurons of the Sw-CPG are mainly distributed in the medulla oblongata, since the supramedullary components are not essential for the generation of swallowing reflex. Previous studies have indicated that these neurons are predominantly located in the nucleus tractus solitarius (NTS) and in the medullary reticular formation (RF) [24–26].

The functional role of neurons in the Sw-CPG has been proposed by Jean [26]: the neurons in and around the NTS are involved in the swallowing rhythm generation, and the neurons in the ventral part of the RF convey signals representing the swallowing movements to the cranial motoneurons. On the other hand, Broussard et al. [27] have advocated the predictive theory regarding the Sw-CPG that the neurons in the interstitial (NTS is) and intermediate subnuclei of the NTS (NTS int) that can receive inputs from upper airway tract have direct projections to the semicompact formation of the NA, which includes pharyngeal and laryngeal motoneurons, contributing to the pharyngeal stage of swallowing, whereas the cells in the central subnucleus of the NTS, to which the NTS int and NTS is neurons can project, send axons to the compact formation of the NA, which includes esophageal motoneurons, contributing to the esophageal stage.

To reveal the neuronal activity and morphology of the Sw-CPG, we recorded and labeled the swallowing-related neurons (SRNs), whose activity changed during fictive swallowing evoked by electrical stimulation of the SLN, in the medulla oblongata in anesthetized paralyzed guinea pigs [28]. Fictive swallowing was identified by bursting activity of the recurrent laryngeal nerve (RLN), the thyrohyoid branch of the hypoglossal nerve (Th-XII), or the pharyngeal branch of the vagal nerve (Ph-X), which corresponds to the pharyngeal stage of swallowing (◘ Fig. 14.5a). The activity of SRNs was classified by three types: (1) early neurons, which fired during the RLN burst corresponding to the pharyngeal stage of swallowing, (2) late neurons that were activated after the RLN burst presumably corresponding to the esophageal stage, and (3) inhibited neurons, whose activity ceased during swallowing (◘ Fig. 14.5b). Our results also indicated that these SRNs were broadly distributed in the NTS and RF, and

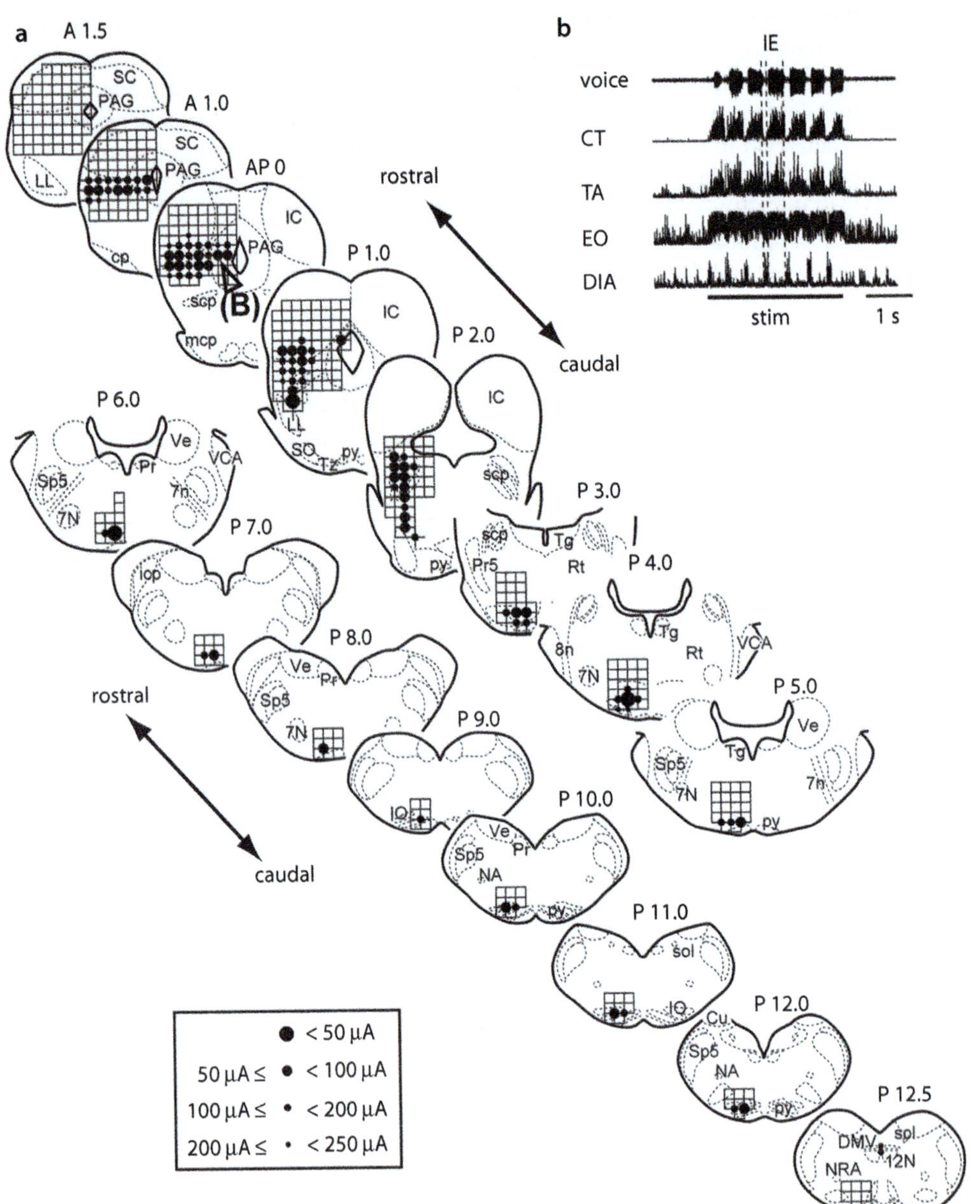

Fig. 14.2 (**a**) Locations of brainstem sites where electrical stimulation evoked vocalization in a guinea pig are shown in transverse sections. The vocal sites are indicated by *closed circles* on the intersections of grid lattices (in 0.5 mm) where electrical stimuli were delivered. The threshold current for evoking vocalization is depicted as *circle* diameter. *Cp* cerebral peduncle, *Cu* cuneate nucleus, *IC* inferior colliculus, *icp* inferior cerebellar peduncle, *IO* inferior olive, *LL* lateral lemniscus, *mcp* middle cerebellar peduncle, *PAG* periaqueductal gray, *Pr* prepositus nucleus, *Pr5* principal sensory trigeminal nucleus, *py* pyramidal tract, *Rt* reticular nucleus, *SC* superior colliculus, *scp* superior cerebellar peduncle, *SO* superior olive, *Sp5* spinal trigeminal tract, *Tg* tegmental nucleus, *Tz* trapezoid body, *VCA* ventral cochlear nucleus, *Ve* vestibular nucleus, *7n* facial nerve, *8n* vestibulocochlear nerve, *DMV* dorsal motor nucleus of the vagus, *12N* hypoglossal nucleus. (**b**) Representative activities of the laryngeal and respiratory muscles during electrical stimulation to the call sites. This vocal-related muscle activity was recorded at the site designated by *open arrowhead* in the "PAG-medulla call area." Vocalization is characterized by the activation of the diaphragm (DIA) [inspiratory phase (I)] followed by the bursting activity of the thyroarytenoid (TA), cricothyroid (CT), and external oblique (EO) muscles [expiratory phase (E)]. Periods of stimulation are indicated by thick lines at the bottom of electromyographic records. stim: duration of electrical stimulation to the call site (From Ref. [18])

their axonal projections represented part of complex neuronal circuitry (Fig. 14.6). As shown in this study, almost all of the SRNs in the NTS had axonal collaterals to the NTS, which suggests that there is the neuronal circuit within the NTS such as the dorsal swallowing group proposed by Jean [26]. Otherwise, the SRNs in the NTS and RF often projected to each other's area, whereas some neurons in the NTS and RF sent axon to the cranial motor nuclei including the NA and hypoglossal nuclei. In addition, some neurons in the RF projected to the other side of the brainstem. In conclusion, we proposed the probable neuronal circuitry involved in swallowing: the SRNs could constitute the local neuronal

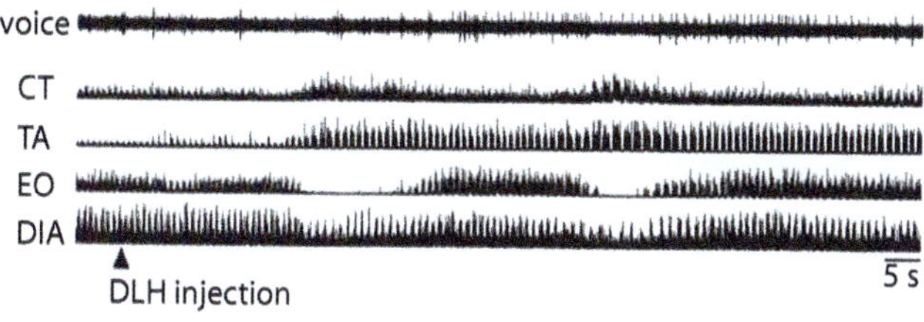

Fig. 14.3 Changes in laryngeal and respiratory muscle activities after excitatory amino acid [D,L-homocysteic acid (DLH)] injection in the vicinity of the nucleus retroambiguus (NRA). DLH injection increased TA and CT muscle activity but did not produce vocalization. The initiation of DLH injection is indicated by an *arrowhead* (From Ref. [18])

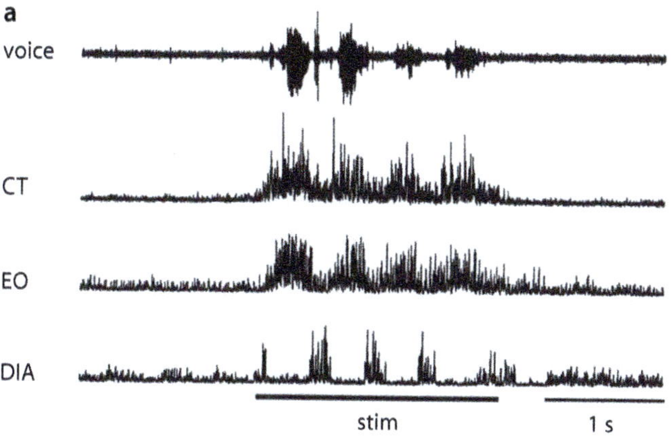

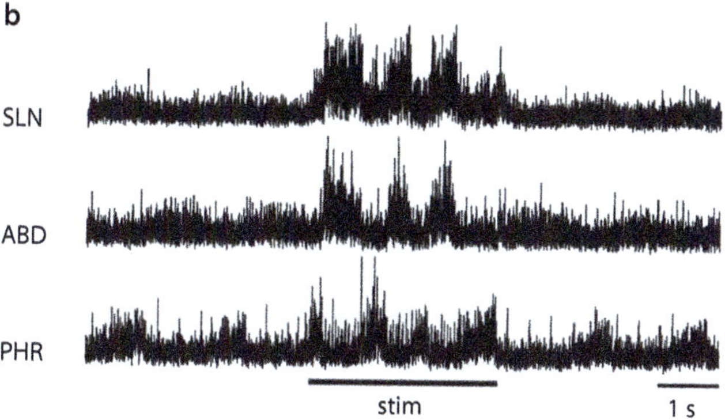

Fig. 14.4 The motor pattern of the laryngeal and respiratory muscle (**a**) and nerve (**b**) activities during PAG-induced vocalization before (**a**) and after (**b**) paralyzation, respectively. The data in **a** and **b** were obtained in the same animal. Period of call site stimulation (stim) is indicated by *thick line* at the bottom of each panel. *SLN* the external branch of the superior laryngeal nerve, *ABD* the L1 abdominal muscle nerve, *PHR* the C5 phrenic nerve (From Ref. [18])

circuits within the NTS that may contribute to the swallowing rhythm generation, the reciprocal connections between the NTS and RF that may shape the motor outputs, the bilateral interconnections in the RF that may synchronize the swallowing outputs, the connections from the NTS and RF to the cranial motor nuclei involved in swallowing that may act as the premotor neurons, and the motoneurons that may integrate the swallowing motor outputs (Fig. 14.7). Further studies will be necessary to understand the network mechanisms involved in swallowing. For example, if the intrinsic properties of the SRNs could be investigated, the more detailed network properties responsible for the generation of this well-coordinated motor sequence would be revealed.

14.4 Multifunctional Respiratory Neurons in Relation to the Laryngeal Movements

The larynx plays a crucial role in voice production, the airway defensive reflexes including swallowing and coughing, as well as respiration [29–33]. In addition, vocalization and these airway defensive reflexes are generated by contractions of the respiratory and upper airway muscles, whose motor actions can take in oxygen and release carbon dioxide in the lung during breathing. These non-respiratory behaviors are thus required by modification of normal respiratory rhythm. The phenomenon that respiratory rhythm is altered in synchrony with these behaviors suggests that the neuronal networks responsible for respiration and those non-respiratory behaviors are overlapped and therefore the respiratory CPG can be shared among the CPGs of those non-respiratory behaviors.

To determine whether the respiratory neurons are included among the CPGs of those behaviors, we compared the activity of the respiratory neurons during breathing with that during those non-respiratory behaviors such as vocalization, swallowing, and coughing in anesthetized paralyzed guinea pigs [34].

Respiratory rhythmogenesis is thought to be regulated by the brainstem neural network, consisting of the DRG, the longitudinal column from pFRG/RTN to cVRG, and PRG, as described above (Fig. 14.1) [7]. We focused on the respiratory neurons located between the BötC and rVRG.

To elucidate the neuronal activity during respiratory and non-respiratory behaviors, we recorded the extracellular activity of the respiratory neurons during fictive respiration, vocalization, swallowing, and coughing. To evoke fictive vocalization, we delivered electrical stimulation to the PAG or the pontine call site in the dorsal pontine tegmentum (Fig. 14.8a) [18, 35]. Fictive swallowing was elicited by electrical stimulation of the SLN (Fig. 14.8b) [28, 36]. Fictive coughing was evoked by mechanical irritation of tracheal mucosa or by electrical stimulation of the RLN and identified by bursting activity of the RLN and ABD preceded by PHR activity (Fig. 14.8c) [37, 38].

We recorded three types of respiratory neurons in the rostral ventrolateral medulla: expiratory, inspiratory, and phase-spanning neurons (Fig. 14.9). The expiratory and inspiratory neurons were additionally characterized regarding their firing rate trajectories: augmenting (AUG), decrementing (DEC), and constant (CON) firing patterns. The phase-spanning neurons were subdivided into the inspiratory-expiratory (IE) and expiratory-inspiratory (EI) neurons.

The specific tendency of firing pattern was observed for each type of the respiratory neurons during the non-

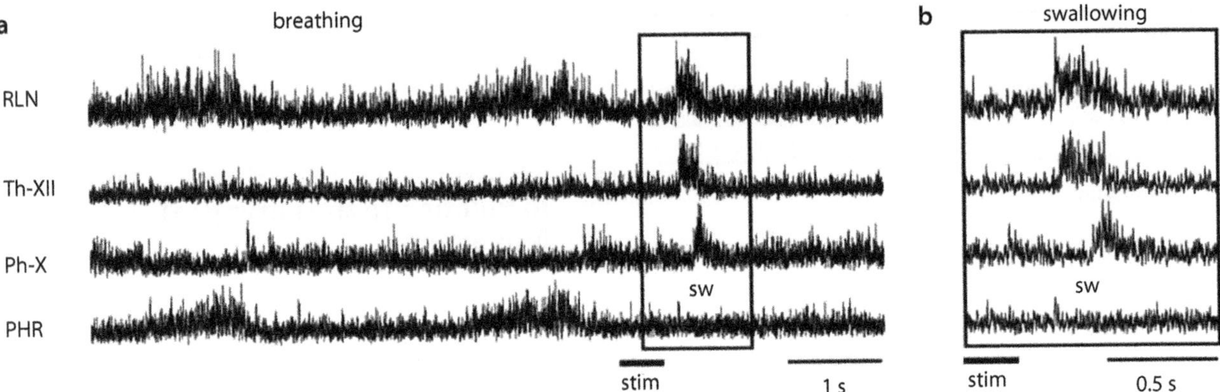
a
breathing
RLN
Th-XII
Ph-X
PHR
sw
stim
1 s
b
swallowing
sw
stim
0.5 s

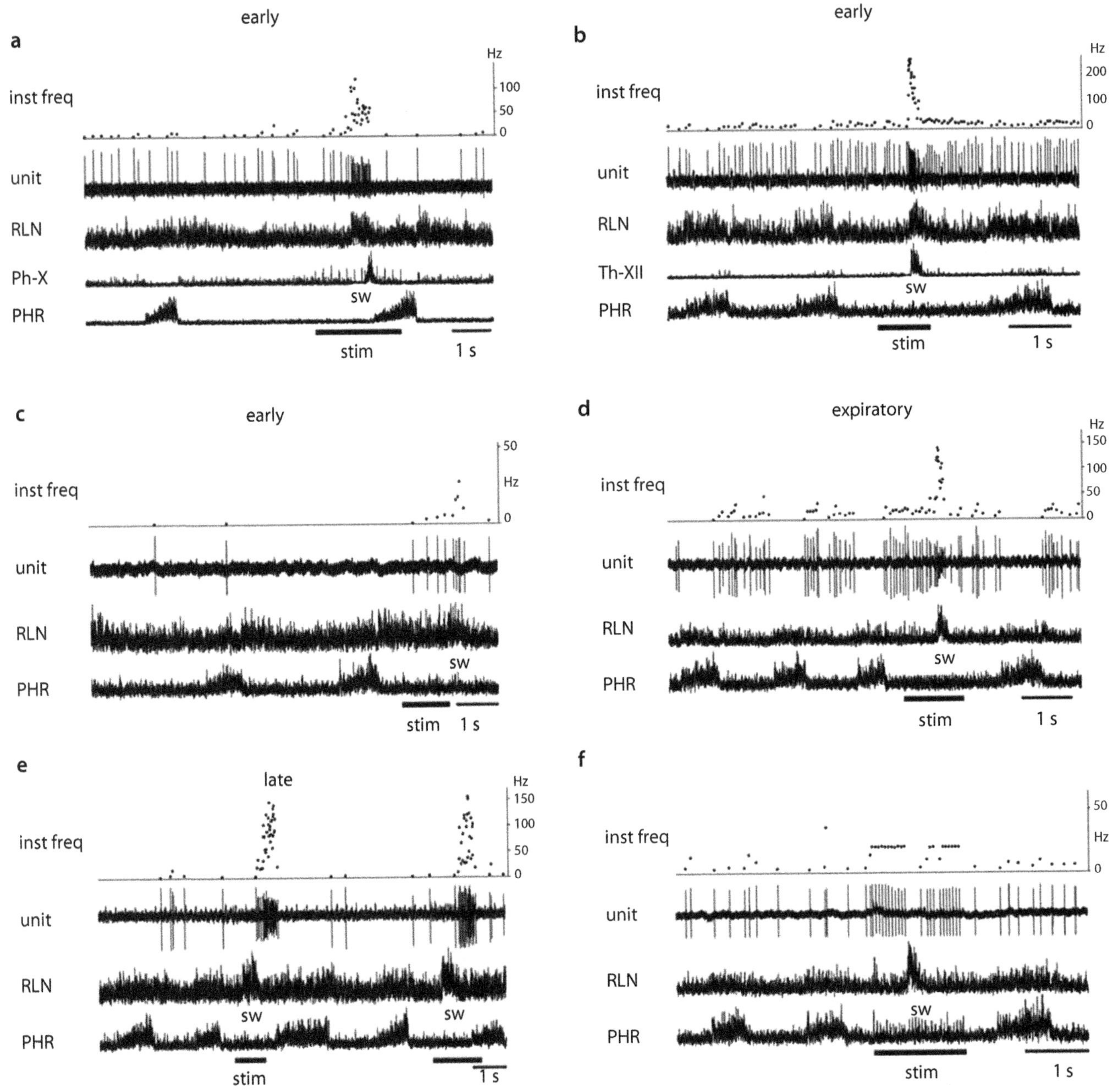

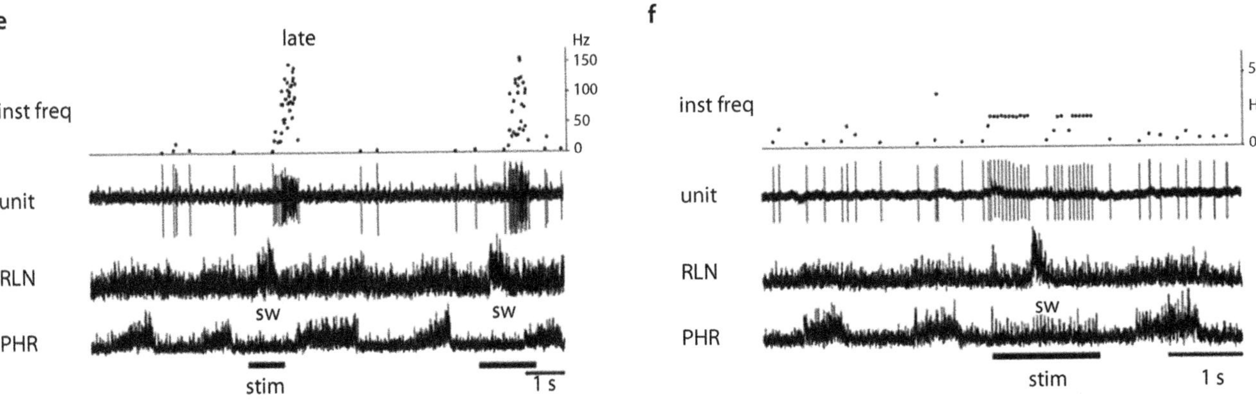
a
early
Hz
inst freq
100
50
0
unit
RLN
Ph-X
PHR
sw
stim
1 s
b
early
Hz
200
100
0
inst freq
unit
RLN
Th-XII
PHR
sw
stim
1 s
c
early
50
Hz
0
inst freq
unit
RLN
PHR
sw
stim
1 s
d
expiratory
Hz
150
100
50
0
inst freq
unit
RLN
PHR
sw
stim
1 s
e
late
Hz
150
100
50
0
inst freq
unit
RLN
PHR
sw
sw
stim
1 s
f
50
Hz
0
inst freq
unit
RLN
PHR
sw
stim
1 s

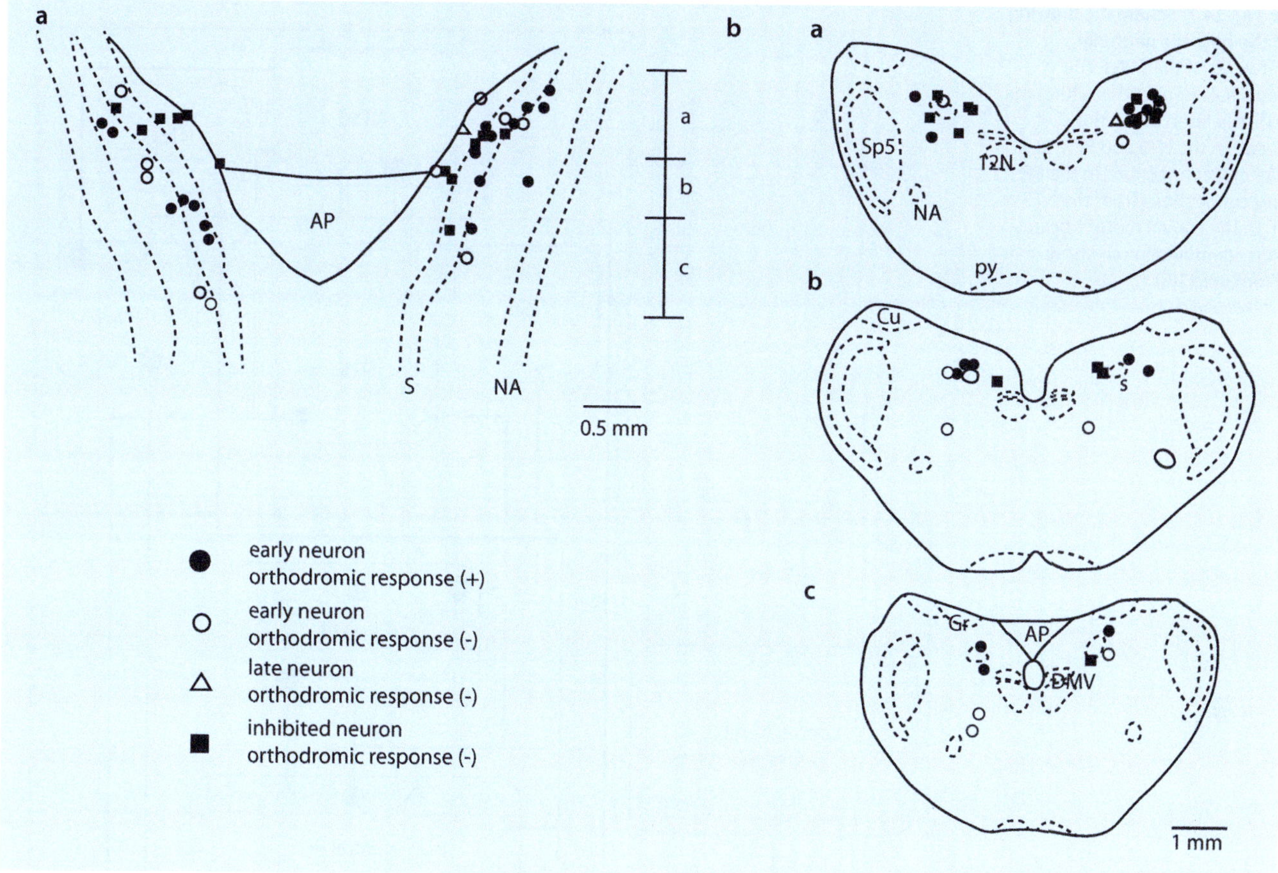

Fig. 14.6 Locations of SRNs recorded in our study. Letters beside the horizontal section (**a**) show the anterior-posterior region represented by each of the transverse sections (**b**). *Circles*, *triangles*, and *squares* represent locations of early-, late-, and inhibited-type neurons, respectively. *Closed* and *open symbols* represent neurons that did and did not respond orthodromically to single-shock stimulation of the SLN, respectively. *AP* area postrema, *Gr* gracile nucleus (From Ref. [28])

Fig. 14.5 Motor patterns of fictive breathing (**a**-(*a*)) and swallowing (**a**-(*b*)). Fictive swallowing was identified by bursting activities of the recurrent laryngeal nerve (RLN), pharyngeal branch of the vagus nerve (Ph-X), and the thyrohyoid muscle branch of the hypoglossal nerve (Th-XII) evoked by stimulation of the superior laryngeal nerve (SLN). High-speed recordings in the period indicated by the *rectangular box* in (**a**-(*a*)) are shown in (**a**-(*b*)). The pharyngeal stage of swallowing began with the bursts of the RLN and Th-XII, whereas the Ph-X burst lagged behind in time of onset. Duration of SLN stimulation (stim) is indicated by the *horizontal bars* at the bottom. Firing patterns of swallowing-related neurons (SRNs) (**b**), including early (**b**-(*a*) to **b**-(*d*)), late (**b**-(*e*)), and inhibited (**b**-(*f*)) neurons. Early neurons fired during the whole pharyngeal stage (**b**-(*a*)), during its early part (**b**-(*b*)), and during its latter part (**b**-(*c*)), respectively. The expiratory-related neuron in panel **b**-(*d*) was activated during the RLN burst. Meanwhile, the late neuron in panel **b**-(*e*) was activated after the swallowing-related RLN burst corresponding to the esophageal stage. The inhibited neuron in panel **b**-(*f*) stopped firing during the pharyngeal stage. *Inst freq.* instantaneous frequency (From Ref. [28])

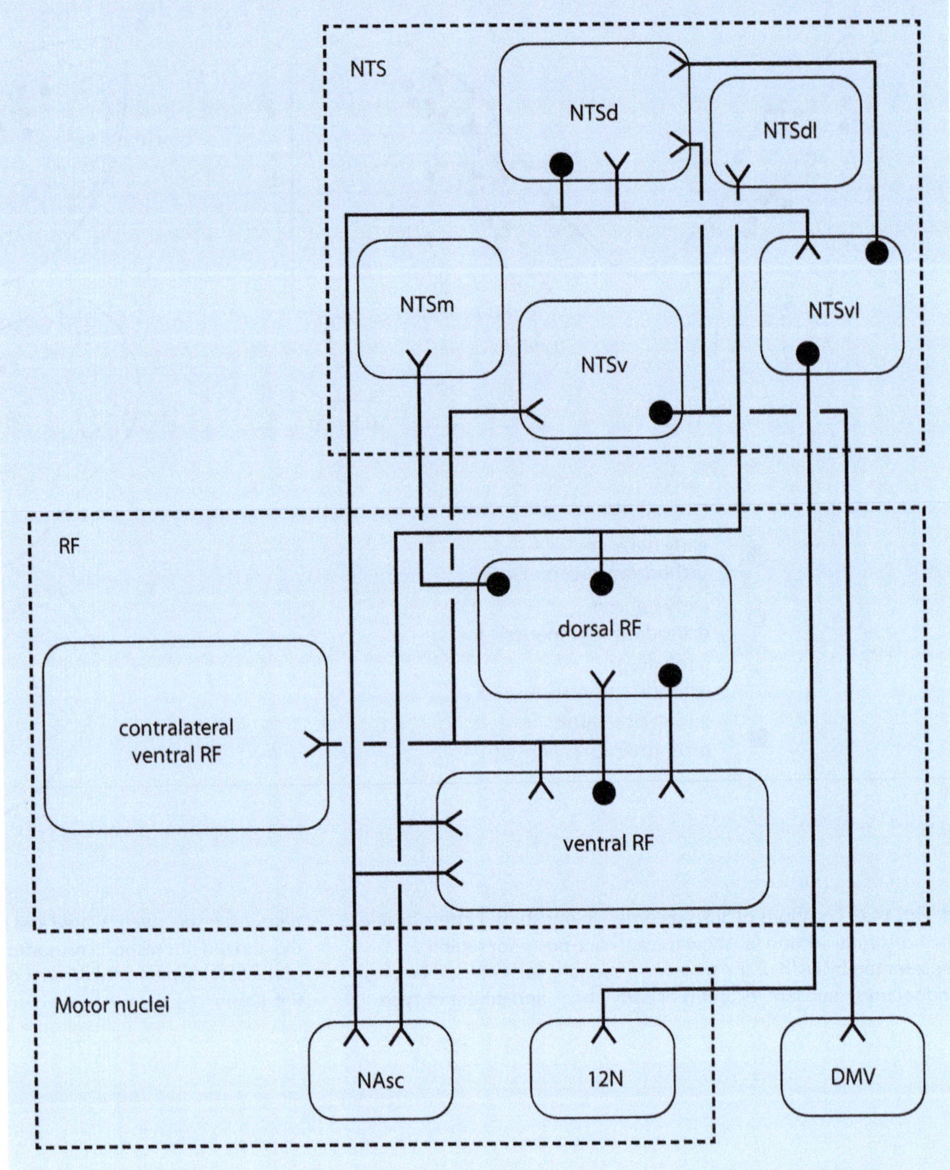

Fig. 14.7 Schematic drawing of the possible neuronal networks of the SRNs. The neuronal connections within the NTS, the interconnections between the NTS and RF, the bilateral connections in the RF, and connections from the NTS or RF to the cranial motor nucleus were identified in our study (From Ref. [28])

respiratory behaviors in this study. The E-AUG neurons in the BötC whose activity can suppress the upper airway motoneuronal activity were generally silent during vocalization, swallowing, and the compressive phase of coughing (Fig. 14.10) [39–42]. This inactivation may facilitate the activity of laryngeal motoneurons during these behaviors. Many E-DEC neurons in the rVRG were activated during all behaviors tested, some of which are possibly upper airway respiratory motoneurons including laryngeal motoneurons (Fig. 14.11) [8, 43–47]. Many E-CON neurons were activated during vocalization and coughing, but did not discharge during swallowing. Some vocal-inactive E-AUG and E-CON neurons resumed firing when the vocal activity was attenuated at the last part of the stimulus-induced expiration (Figs. 14.10a and 14.12). Although their functional role has not been declared, the cells may play a role in the termination of vocalization. The I-AUG neurons, broadly distributed in the rVRG, were typically activated in synchrony with the phrenic discharge during vocalization and coughing [47]. On the contrary, some "late-inspiratory neurons" discharged during the expiratory phase of coughing, probably contributing to the inspiratory-expiratory phase transition or acting as the pharyngeal motoneurons during coughing (Fig. 14.13) [48, 49]. Some I-AUG neurons fired during the period of "swallow-breath," suggesting that these neurons, which could be the phrenic premotor neurons, participate in the generation of "swallow-breath" [47]. The discharge patterns of I-DEC neurons remained unchanged during the inspiratory phase of vocalization and coughing, while these neurons were silent during swallowing. The

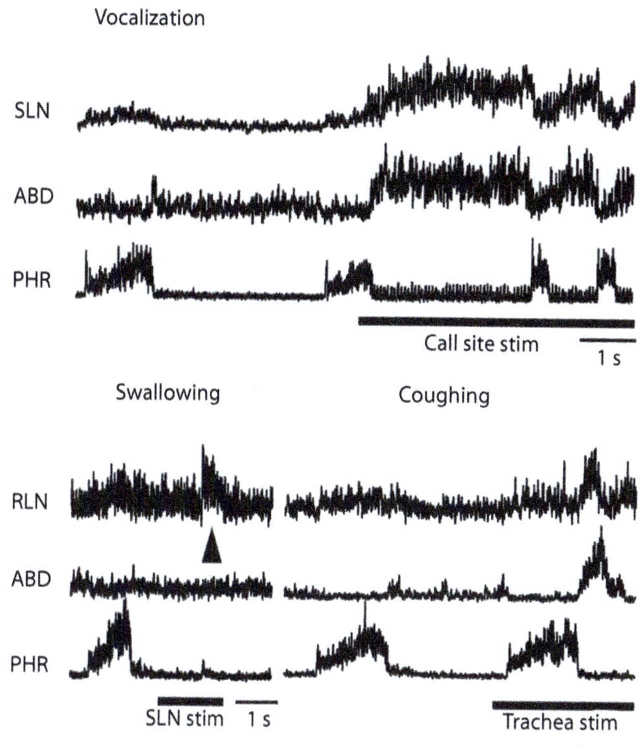

Fig. 14.8 Activities of the efferent nerves innervating the upper airway muscles involved in vocalization (**a**), swallowing (**b**), and coughing (**c**). Fictive vocalization was evoked by electrical stimulation of the periaqueductal gray or pontine call site. The vocal phase was identified by bursting activity of the SLN and the ABD followed by activation of the PHR (**a**). Electrical stimulation of the SLN elicited fictive swallowing identified by bursting activity of the RLN (*arrowhead*) (**b**). Fictive coughing, which was evoked by mechanical stimulation of the trachea, consisted of an abrupt burst of the abdominal nerve accompanied by bursting activity of the RLN following phrenic nerve activation (**c**) (From Ref. [34])

I-CON neurons were activated during the inspiratory phase of vocalization and coughing. Many phase-spanning neurons, which may play a role in the phase transition during respiration, fired during vocalization, swallowing, and coughing (Figs. 14.14 and 14.15) [50–52]. The strong activation of these neurons during the vocal phase may play a key role in the preservation of vocal emission as well as the phase transition, whereas the activation during swallowing may inhibit respiration. On the other hand, the EI neurons, some of which could be the pharyngeal motoneurons, may help to keep the pressure of forceful coughing [49]. However, the connectivity between the phase-spanning neurons and the other brainstem respiratory neurons, including laryngeal motoneurons, remains unknown. Further studies are needed to explore this possibility.

Based on our data, we propose that the respiratory neuronal networks possess the ability to reconfigurate their own networks and that the individual respiratory neuron alters its activity in a specific manner, which is adjustable to provide each non-respiratory behavior. Our data thus support the view that the medullary respiratory neurons are multifunctional and can be shared in the CPGs involved in the non-respiratory laryngeal behaviors.

14.5 Perspectives

While the principal function of the larynx is phylogenetically the airway protection including feeding and expelling the foreign body to prevent airway from aspiration, various laryngeal functions including phonation have been acquired during the course of evolution. Simultaneously, the network organization responsible for these behaviors should have been constructed. Despite the complexity of the CPG networks, it is reasonable that the brainstem neuronal networks serve the efficient and effective processing during these behaviors. To realize this concept, multifunctional neuronal activity may be indispensable. Previous studies have emphasized the importance of the premotor neurons including respiratory neurons that can directly control laryngeal movements, which may have multifunctional properties [27, 41, 53–55]. On the contrary, the behavior-specific neurons, such as the SRNs reported in our study, are likely to play an essential role in the generation of these behaviors. Although these CPG networks are not fully understood, the declaration of both the physiological and anatomical properties of the CPG neurons will improve understanding of the network mechanisms responsible for the laryngeal movements.

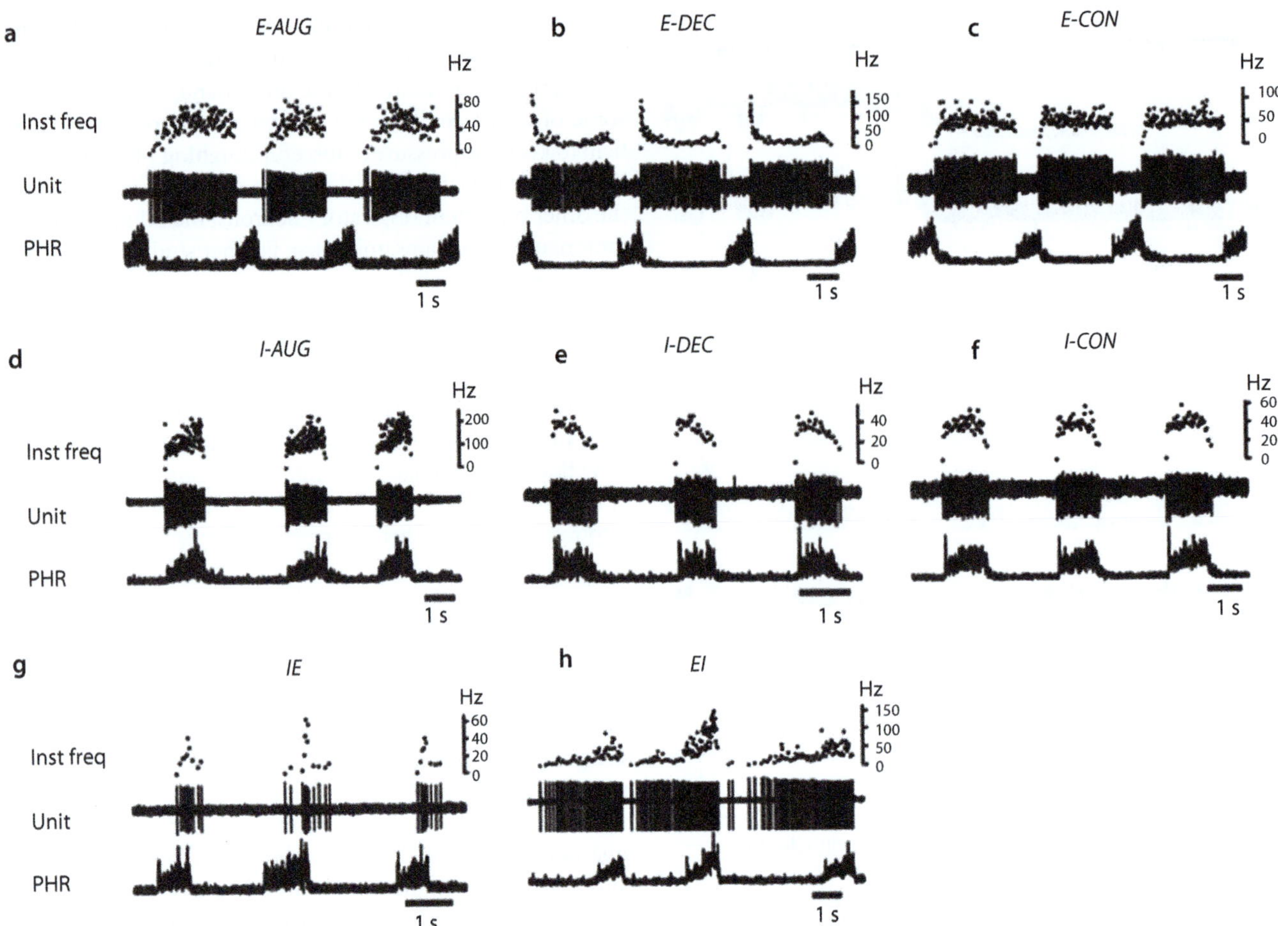

Fig. 14.9 Subtypes of respiratory neurons in the rostral ventrolateral medulla. Expiratory neurons with an augmenting (E-AUG) (**a**), decrementing (E-DEC) (**b**), and constant (E-CON) (**c**) firing patterns, exhibiting a gradual increase, decrease, and no change in firing rates during the expiratory phase, respectively. Inspiratory neurons with augmenting (I-AUG) (**d**), decrementing (I-DEC) (**e**), and constant (I-CON) (**f**) firing patterns. Panels (**g**) and (**h**) show cell firings with phase-spanning activity which began during inspiration and continued into expiration (inspiration to expiration phase spanning, IE) and began during expiration and continued into inspiration (expiration to inspiration phase spanning, EI), respectively (From Ref. [34])

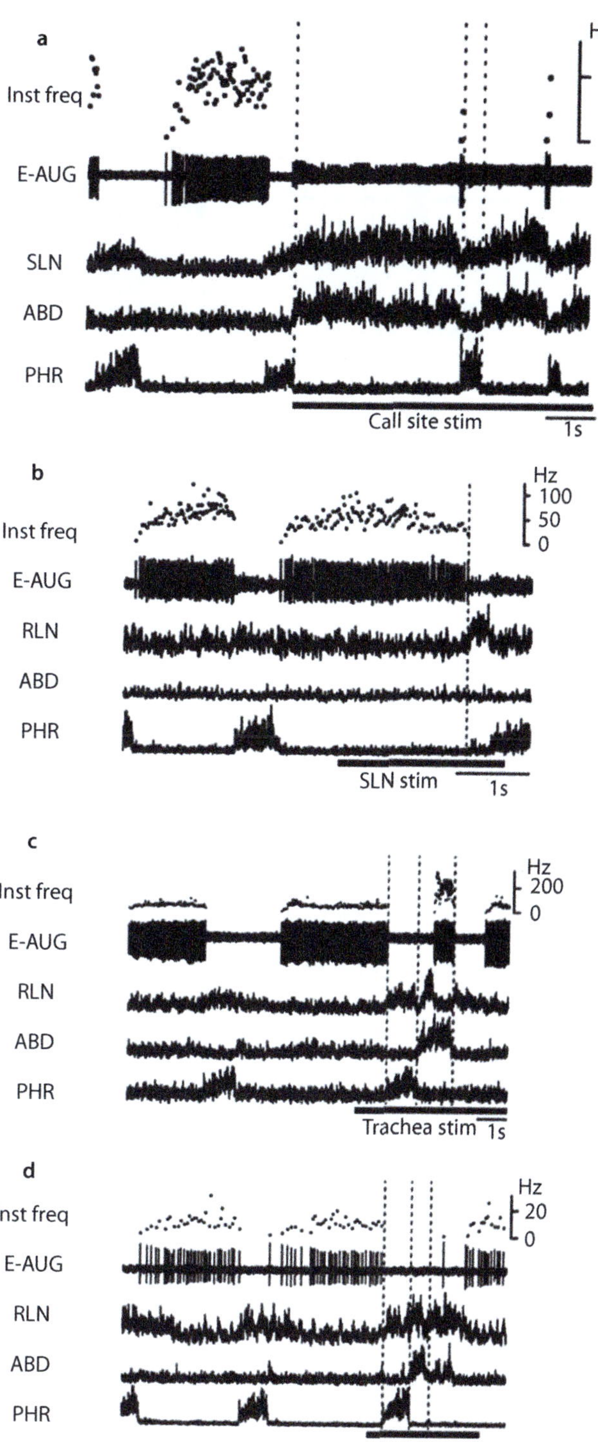

Figure 14.10 Representative firing patterns of the E-AUG neurons during vocalization (**a**), swallowing (**b**), and coughing (**c**, **d**). The vocalization-inactive E-AUG neuron in panel **a** was silent during the period of SLN and ABD bursts corresponding to the vocal phase. The E-AUG neuron in panel **b** was silent during swallowing identified by the swallow-related RLN burst induced by SLN stimulation. The E-AUG neuron in panel **c** fired just after the bursting activity of the RLN during the expiratory phase of coughing presumably corresponding to the expulsive phase of coughing. The E-AUG neuron in panel **d** was silent during fictive coughing. *Thick line* at the bottom of each panel represents the stimulus duration of the call site, SLN, RLN, or tracheal mucosa (call site stim, SLN stim, RLN stim, or trachea stim). *Dashed lines* indicate the respiratory phase transitions of vocalization (**a**), coughing (**c**), and the initiation of swallowing (**b**) (Reproduced, with permission, from Ref. [34] (2014))

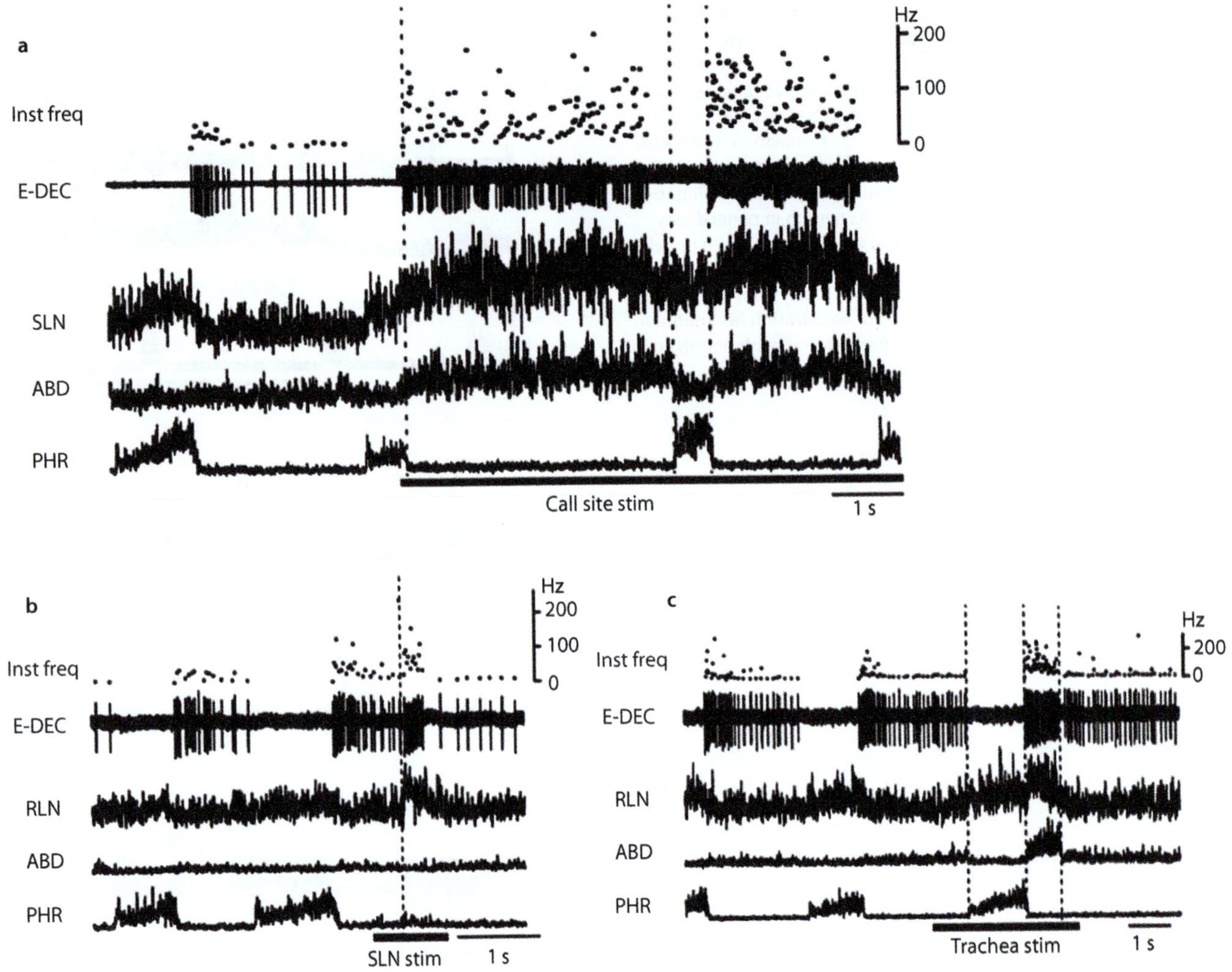

Fig. 14.11 Activity of the E-DEC neurons during vocalization (**a**), swallowing (**b**), and coughing (**c**). The E-DEC neuron in panel **a** showed increased firing rates during vocalization compared to before stimulation. The E-DEC neuron in panel **b** was activated during swallowing. The E-DEC neuron in panel **c** was activated with a decrementing discharge pattern during the expiratory phase of coughing (Reproduced, with permission, from Ref. [34] (2014))

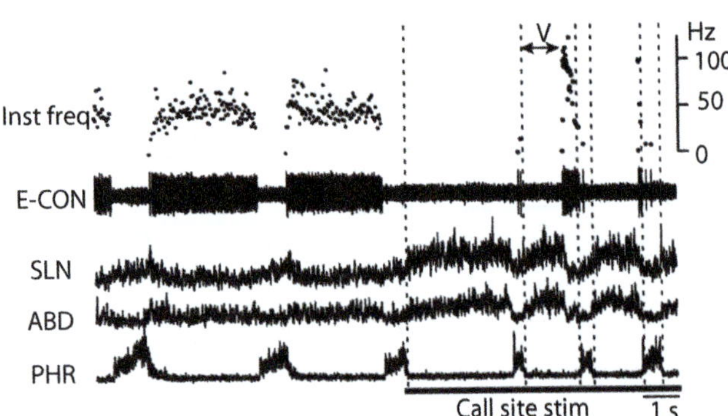

Fig. 14.12 The E-CON neuron was silent during the period of SLN and ABD bursts corresponding to the vocal phase (V), but fired at the end of the stimulus-induced expiration during which the bursts were attenuated (Reproduced, with permission, from Ref. [34] (2014))

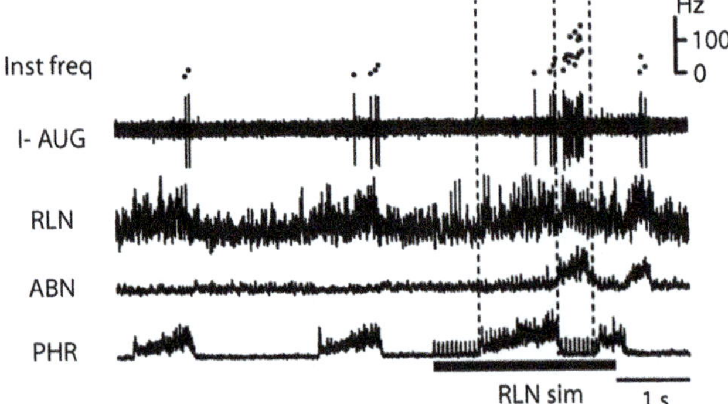

Fig. 14.13 Firing of the inspiratory neurons during coughing. This late-onset I-AUG neuron was activated during the expiratory phase of coughing (Reproduced, with permission, from Ref. [34] (2014))

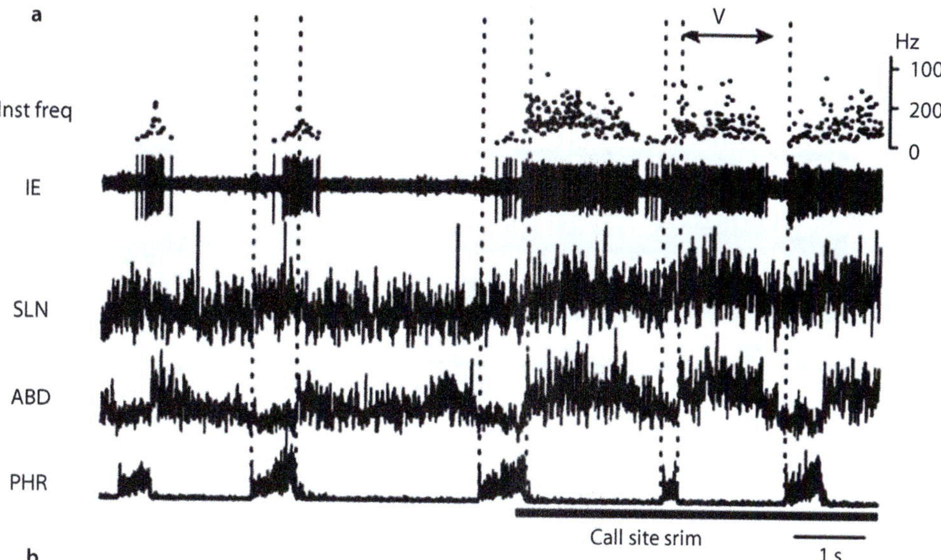

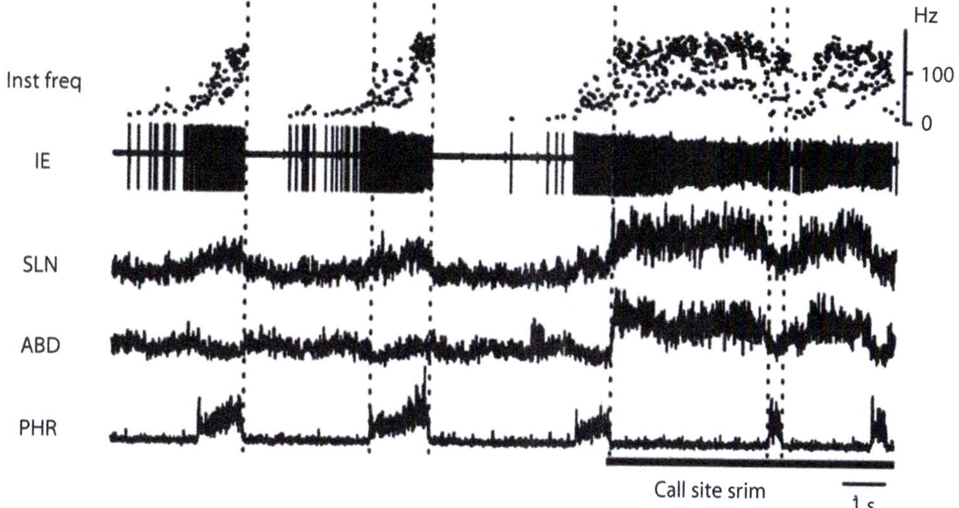

Fig. 14.14 Activity of phase-spanning neurons during vocalization. The IE neuron in panel (**a**) strongly fired during the vocal phase. This neuron sometimes ceased its firing when the vocal-related SLN and ABD bursts were attenuated at the end of the expiratory phase during the call site stimulation. The EI neuron in panel (**b**) weakly fired during the late expiration of control respiration, but strongly fired throughout the vocal phase (From Ref. [34])

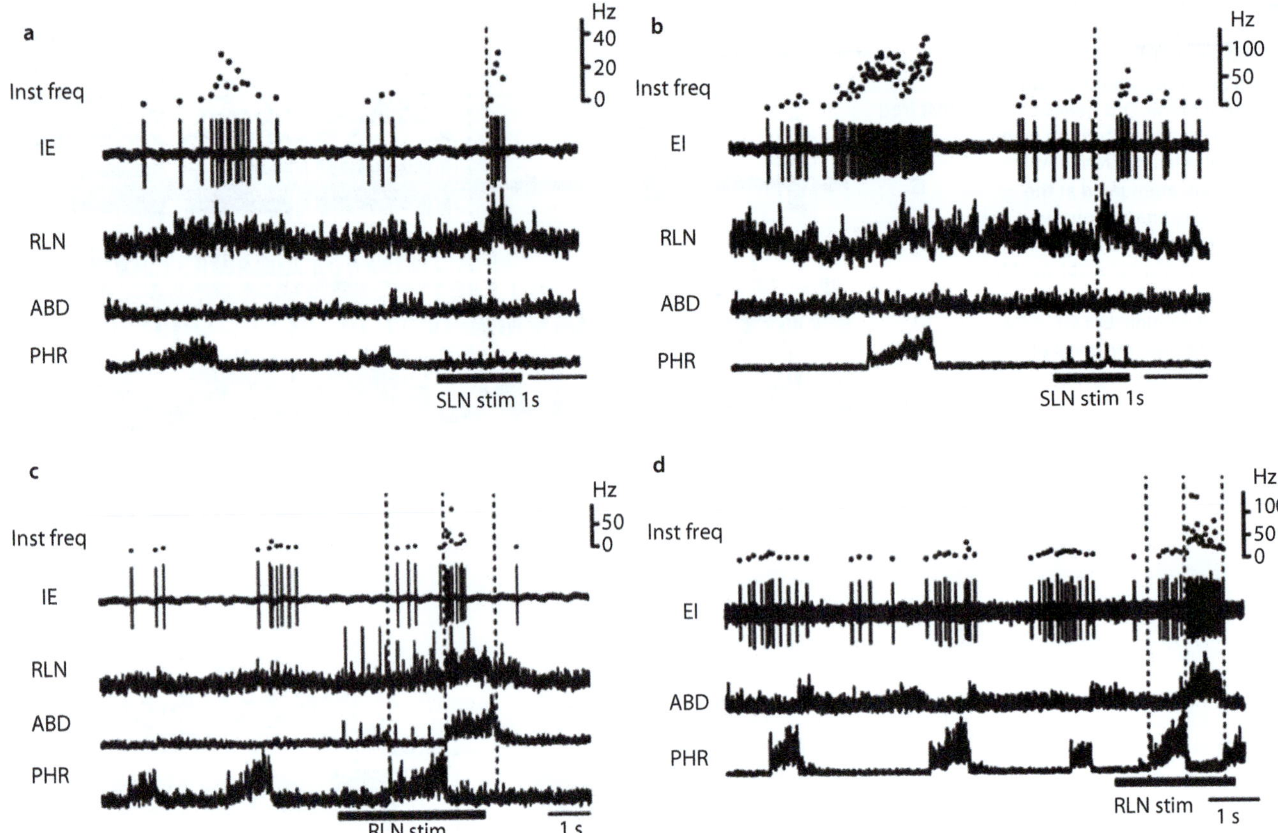

Fig. 14.15 Firing patterns of phase-spanning neurons during swallowing (**a**, **b**) and coughing (**c**, **d**). The IE neuron in panel **a** discharged during swallowing. The EI neuron in panel **b** began to fire approximately 0.3 s after the onset of the RLN burst. The IE neuron in panel **c** fired at the onset of the expiratory phase of coughing. The EI neuron in panel **d** strongly discharged during the expiratory phase of coughing (Reproduced, with permission, from Ref. [34] (2014))

References

1. Miller AD, Tan LK, Lakos SF. Brainstem projections to cats' upper lumbar spinal cord: implications for abdominal muscle control. Brain Res. 1989;493(2):348–56.
2. VanderHorst VG, Terasawa E, Ralston 3rd HJ. Monosynaptic projections from the nucleus retroambiguus region to laryngeal motoneurons in the rhesus monkey. Neuroscience. 2001;107(1):117–25.
3. Altschuler SM, Bao X, Bieger D, Hopkins DA, Miselis RR. Viscerotopic representation of the upper alimentary tract in the rat: sensory ganglia and nuclei of the solitary and spinal trigeminal tracts. J Comp Neurol. 1989;283(2):248–68. 1989/05/08 ed.
4. Jürgens U. Neural pathways underlying vocal control. Neurosci Biobehav Rev. 2002;26(2):235–58.
5. Berger AJ, Averill DB, Cameron WE. Morphology of inspiratory neurons located in the ventrolateral nucleus of the tractus solitarius of the cat. J Comp Neurol. 1984;224(1):60–70.
6. Iscoe S, Grélot L, Bianchi AL. Responses of inspiratory neurons of the dorsal respiratory group to stimulation of expiratory muscle and vagal afferents. Brain Res. 1990;507(2):281–8.
7. Feldman JL, Del Negro CA. Looking for inspiration: new perspectives on respiratory rhythm. Nat Rev Neurosci. 2006;7(3):232–42.
8. Bianchi AL, Denavit-Saubié M, Champagnat J. Central control of breathing in mammals: neuronal circuitry, membrane properties, and neurotransmitters. Physiol Rev. 1995;75(1):1–45.
9. Adametz J, O'Leary JL. Experimental mutism resulting from periaqueductal lesions in cats. Neurology. 1959;9:636–42.
10. Botez MI, Barbeau A. Role of subcortical structures, and particulary of the thalamus, in the mechanisms of speech and language. Int J Neurol. 1971;8:300–20.
11. Esposito A, Demeurisse G, Alberti B, Fabbro F. Complete mutism after midbrain periaqueductal gray lesion. Neuroreport. 1999;10:681–5.
12. Kelly AH, Beaton LE, Magoun HW. A midbrain mechanism for facio-vocal activity. J Neurophysiol. 1946;9:181–9.
13. Jürgens U, Pratt R. Role of the periaqueductal grey in vocal expression of emotion. Brain Res. 1979;167(2):367–78.
14. Jürgens U, Richter K. Glutamate-induced vocalization in the squirrel monkey. Brain Res. 1986;373(1–2):349–58.
15. Zhang SP, Davis PJ, Bandler R, Carrive P. Brain stem integration of vocalization: role of the midbrain periaqueductal gray. J Neurophysiol. 1994;72(3):1337–56.
16. Holstege G. Anatomical study of the final common pathway for vocalization in the cat. J Comp Neurol. 1989;284(2):242–52.
17. Shiba K, Umezaki T, Zheng Y, Miller AD. The nucleus retroambigualis controls laryngeal muscle activity during vocalization in the cat. Exp Brain Res. 1997;115:513–9.
18. Sugiyama Y, Shiba K, Nakazawa K, Suzuki T, Hisa Y. Brainstem vocalization area in guinea pigs. Neurosci Res. 2010;66(4):359–65.
19. Syka J, Suta D, Popelar J. Responses to species-specific vocalizations in the auditory cortex of awake and anesthetized guinea pigs. Hear Res. 2005;206(1–2):177–84.
20. Suta D, Kvasnak E, Popelar J, Syka J, Kvaˇ E, Kva E, et al. Representation of species-specific vocalizations in the inferior colliculus of the guinea pig. J Neurophysiol. 2003;90(6):3794–808.

21. Suta D, Popelar J, Kvasnak E, Syka J. Representation of species-specific vocalizations in the medial geniculate body of the guinea pig. Exp Brain Res. 2007;183(3):377–88.
22. Kyuhou S, Gemba H. Two vocalization-related subregions in the midbrain periaqueductal gray of the guinea pig. Neuroreport. 1998;9(7): 1607–10.
23. Behrend O, Schuller G. The central acoustic tract and audio-vocal coupling in the horseshoe bat. Rhinolophus rouxi. Eur J Neurosci. 2000;12(12):4268–80.
24. Kessler JP, Jean A. Identification of the medullary swallowing regions in the rat. Exp Brain Res. 1985;57(2):256–63. 1985/01/01 ed.
25. Umezaki T, Matsuse T, Shin T. Medullary swallowing-related neurons in the anesthetized cat. Neuroreport. 1998;9:1793–8.
26. Jean A. Brain stem control of swallowing: neuronal network and cellular mechanisms. Physiol Rev. 2001;81:929–69.
27. Broussard DL, Lynn RB, Wiedner EB, Altschuler SM. Solitarial premotor neuron projections to the rat esophagus and pharynx: implications for control of swallowing. Gastroenterology. 1998;114(6): 1268–75.
28. Sugiyama Y, Shiba K, Nakazawa K, Suzuki T, Umezaki T, Ezure K, et al. Axonal projections of medullary swallowing neurons in guinea pigs. J Comp Neurol. 2011;519(11):2193–211.
29. Ludlow CL. Central nervous system control of the laryngeal muscles in humans. Respir Physiol Neurobiol. 2005;147:205–22.
30. Jürgens U. The neural control of vocalization in mammals: a review. J Voice. 2009;23(1):1–10. 2008/01/22 ed.
31. Paydarfar D, Gilbert RJ, Poppel CS, Nassab PF. Respiratory phase resetting and airflow changes induced by swallowing in humans. J Physiol. 1995;483(1):273–88.
32. Miller AJ. The neurobiology of swallowing and dysphagia. Dev Disabil Res Rev. 2008;14(2):77–86.
33. Karlsson JA, Sant'Ambrogio G, Widdicombe J. Afferent neural pathways in cough and reflex bronchoconstriction. J Appl Physiol. 1988;65(3):1007–23.
34. Sugiyama Y, Shiba K, Mukudai S, Umezaki T, Hisa Y. Activity of respiratory neurons in the rostral medulla during vocalization, swallowing, and coughing in guinea pigs. Neurosci Res. 2014;80: 17–31.
35. De Lanerolle NC. A pontine call site in the domestic cat: behavior and neural pathways. Neuroscience. 1990;37(1):201–14.
36. Nishino T, Honda Y, Kohchi T, Shirahata M, Yonezawa T. Effects of increasing depth of anaesthesia on phrenic nerve and hypoglossal nerve activity during the swallowing reflex in cats. Br J Anaesth. 1985;57(2):208–13.
37. Bolser D. Fictive cough in the cat. J Appl Physiol. 1991;71:2325–31.
38. Grélot L, Milano S. Diaphragmatic and abdominal muscle activity during coughing in the decerebrate cat. Neuroreport. 1991;2(4): 165–8.
39. Jiang C, Lipski J. Extensive monosynaptic inhibition of ventral respiratory group neurons by augmenting neurons in the Bötzinger complex in the cat. Exp Brain Res. 1990;81(3):639–48.
40. Ono K, Shiba K, Nakazawa K, Shimoyama I. Synaptic origin of the respiratory-modulated activity of laryngeal motoneurons. Neuroscience. 2006;140(3):1079–88.
41. Shiba K, Nakazawa K, Ono K, Umezaki T. Multifunctional laryngeal premotor neurons: their activities during breathing, coughing, sneezing, and swallowing. J Neurosci. 2007;27(19):5156–62.
42. Sakamoto T, Katada A, Nonaka S, Takakusaki K. Activities of expiratory neurones of the Bötzinger complex during vocalization in decerebrate cats. Neuroreport. 1996;7:2353–6.
43. Ezure K. Synaptic connections between medullary respiratory neurons and considerations on the genesis of respiratory rhythm. Prog Neurobiol. 1990;35(6):429–50.
44. Zheng Y, Barillot JC, Bianchi AL. Are the post-inspiratory neurons in the decerebrate rat cranial motoneurons or interneurons? Brain Res. 1991;551(1–2):256–66.
45. Nonaka S, Katada A, Sakamoto T, Unno T. Brain stem neural mechanisms for vocalization in decerebrate cats. Ann Otol Rhinol Laryngol Suppl. 1999;108:15–24.
46. Sakamoto T, Nonaka S, Katada A. Control of respiratory muscles during speech and vocalization. In: Miller AD, Bianchi AL, Bishop BP, editors. Neural control of the respiratory muscles. Florida: CRC Press; 1996. p. 249–58.
47. Oku Y, Tanaka I, Ezure K. Activity of bulbar respiratory neurons during fictive coughing and swallowing in the decerebrate cat. J Physiol. 1994;480(Pt 2):309–24.
48. Bianchi AL, Gestreau C. The brainstem respiratory network: an overview of a half century of research. Respir Physiol Neurobiol. 2009;168(1–2):4–12. 2009/05/02 ed.
49. Grélot L, Barillot JC, Bianchi AL. Pharyngeal motoneurones: respiratory-related activity and responses to laryngeal afferents in the decerebrate cat. Exp Brain Res. 1989;78(2):336–44.
50. Cohen MI. Discharge patterns of brain-stem respiratory neurons during Hering-Breuer reflex evoked by lung inflation. J Neurophysiol. 1969;32(3):356–74.
51. Von Euler C. Brain stem mechanisms for generation and control of breathing pattern. In: Cherniack NS, Widdicombe JG, editors. Handbook of physiology: section 3, the respiratory system. Bethesda: Americal Physiological Society; 1986. p. 1–67.
52. Schwarzacher SW, Smith JC, Richter DW. Pre-Bötzinger complex in the cat. J Neurophysiol. 1995;73(4):1452–61.
53. Kuna ST, Remmers JE. Premotor input to hypoglossal motoneurons from Kolliker-Fuse neurons in decerebrate cats. Respir Physiol. 1999;117(2–3):85–95.
54. Ono T, Ishiwata Y, Inaba N, Kuroda T, Nakamura Y. Modulation of the inspiratory-related activity of hypoglossal premotor neurons during ingestion and rejection in the decerebrate cat. J Neurophysiol. 1998;80:48–58.
55. Gestreau C, Dutschmann M, Obled S, Bianchi AL. Activation of XII motoneurons and premotor neurons during various oropharyngeal behaviors. Respir Physiol Neurobiol. 2005;147(2–3):159–76.
56. Berman ALI. The brain stem of the cat. Madison: University of Wsconsin Press; 1968.

The manufacturer's authorised representative in the EU is Springer Nature Customer Service Centre GmbH, Europaplatz 3, 69115 Heidelberg, Germany. If you have any concerns regarding our products, please contact ProductSafety@springernature.com

Printed and bound by CPI Group (UK) Ltd, Croydon, CR0 4YY
15/07/2026
02167660-0001